Introduction to GIS for the Petroleum Industry

Introduction to GIS for the Petroleum Industry

Dean E. Gaddy

PennWell Corporation
1421 South Sheridan Road/P O Box 1260
Tulsa, Oklahoma 74101

800.752.9764
+1.918.831.9421
sales@pennwell.com
www.pennwell-store.com
www.pennwell.com

Managing Editor: Marla Patterson
Production Editor: Sue Rhodes Dodd
Book design: Wes Rowell
Cover design: Matt Berkenbile

Library of Congress Cataloging-in-Publication Data

Gaddy, Dean E.

Introduction to GIS for the petroleum industry / Dean E. Gaddy.– 1st American ed.

p. cm.

ISBN 0-87814-804-3

1. Geographic information systems. 2. Petroleum industry and trade–Data processing.
I. Title.

G70.212 .G34 2003

910'.285–dc22

2003016042

Printed in the United States of America

1 2 3 4 5 07 06 05 04 03

To those who are closest to me:
My wife Polina, my daughter Natalie, and
my very best of friends, Pyotr and Larisa.

Contents

Foreword

Introduction to GIS for the Petroleum Industry was born of a need—a need for a broad work on the maturing field of geographic information system (GIS) technology and the incredible potential it offers to the petroleum industry. GIS encompasses many fields, including computer science, geography, cartography, information management, telecommunications, geodesy, photogrammetry, and remote sensing. The technology is flavored with applications of engineering, network analysis, subsurface modeling, asset management, and many, many others.

GIS has become one of today's leading technologies because it offers an important means of understanding and dealing with some of the most challenging issues of our time. These issues include homeland security, natural resource management, disease outbreaks, and population growth, to name a few.

As Dean Gaddy clearly conveys in this book, GIS technology can help us organize the data about such problems and understand their spatial relationships. It provides a powerful means for analyzing and creating information about those relationships. I look forward to the day when GIS technology will be part of the support process of decision makers who affect our natural and manmade environments—from local areas to those on a global scale. For that day to come, GIS technology must become more widely understood and applied. This book can play an important part in that process for the petroleum industry.

This book is a user's introduction to GIS technology. Some knowledge of computer and data processing concepts is helpful but not essential in understanding it. Explanations of basic concepts and terms make understandable to the layperson the technical issues that affect GIS development, such as accuracy, scales, and projections. The book provides information and perspectives on the major requirements and issues surrounding GIS technology. These include understanding the various spatial models, choosing a specific application to solve a particular problem, managing the quality of data, designing and projecting maps, and others.

A unique aspect of this book is that it is intended for both users and managers involved in all aspects of the business, ranging from geologic exploration to oil sales. It assumes that the reader has no previous background in the field of GIS. The content is organized to provide the reader with a logically complete overview of the knowledge needed to make informed decisions about the applicability of the technology.

The chapters are organized in a sequence of steps fundamental to the planning, implementation, and operation of a GIS program. For these individuals, this book should provide real value. In addition, managers will find that the author has provided a broad overview of GIS in the petroleum industry, allowing them to read through the book in its entirety or use it as a quick reference.

The book discusses concerns critical to managers and validates the value proposition GIS technology offers. The text is supplemented with a large variety of useful graphics and illustrations, which readers will want to return to time and time again. (Publisher's note: Color plates for each chapter are located at the center of the book.) Actual GIS projects and related maps and reports demonstrating GIS petroleum applications are cited extensively throughout the book.

This well-written book is comprehensive, thorough, and valuable to persons in the field. In its treatment of a broad range of topics, the book provides prudent advice offered from the perspective of experience. It also describes the problems and limitations of GIS technology, while providing insights to the future of GIS.

Dean Gaddy has written an outstanding book about GIS and its application in the petroleum industry. If it is widely read and applied, I believe it can make a difference in nearly every business aspect of the petroleum industry.

David DiSera, President
Geospatial Information & Technology Association

Acknowledgments

The author wishes to acknowledge the following individuals and organizations for their contributions to this book. These contributions include case studies, copyright permissions, and reviews of the written manuscript and artwork:

Joseph K. Berry, for his review of certain sections within chapter 3 and his excellent insights into GIS in general.

John Grace of Earth Science Associates Inc., for his case-study contributions to chapter 8, his review of chapter 3, and his devotion to GIS and the petroleum industry.

Daniel Johnston and the *Oil & Gas Journal* On-line Research Center, for providing data on the Nelson Complexity Index.

Martin Rayon of Quest, for his contributions and review of chapters 4 and 5 and his excellent tutorials on geodesy.

John Brand with Geoscience Earth and Marine Services Inc., for submitting to several interviews on pipeline corridor selection (chapter 2).

Chris Shill, formerly with Arco, for his GIS-based marketing innovations (chapter 2).

Steve Adam of Canadian Geomatic Solutions Ltd., for his contributions to chapter 7 and unique insights into state-of-the-art satellite imagery.

Gary Napier of Satellite Imaging, for providing amazingly high-resolution IKONOS imagery.

Floyd Sabins, a remote-sensing veteran, for providing permission to publish several Landsat figures (chapter 7).

Professor Shattri Mansor of the University of Putra Malaysia, for his oil-spill case study discussed in chapter 7.

Ted Mirenda of PennEnergy Data, for his contributions to chapter 8 and a look behind the mechanics of a web-based GIS.

Robert G. Davis and Keith Fraley with Spud IT LLC, for their unique GIS/GPS contributions to drilling (chapter 6).

Special mention goes to Petroleum Argus's Moscow office for providing data on Russian refinery output.

Wood MacKenzie for providing data on Russian infrastructure.

Elizabeth Roberts of GITA for her assistance in acquiring several excellent references.

The editors of *GPS World* for their well-documented history of GPS technology.

The editors of the *Oil & Gas Journal*, including Alan Petzet, Bob Tippee, Warren True, Guntis Moritis, and Leo Aalund.

The international operations management team at Devon Energy Corp. consisting of Danny Nasser, Rick Mitchell, and Earl Reynolds, for their unwavering support.

For those who contributed to this book but were regrettably not mentioned, I sincerely thank you for your help. This book would not have been possible without you.

1

An Introduction to GIS

Until now, the petroleum industry has only used bits and pieces of GIS (geographical information system). For sure, oilfield programmers have developed amazing programs that surpass other industrial applications in terms of geospatial wizardry. Earth scientists, for example, regularly employ three-dimensional (3-D) models that link interpretive insights and geophysical data to generate graphically interactive cross sections and maps.

Some of these programs, used in tandem with reservoir data, can even process time-lapse (four-dimensional or 4-D) shots of petroleum reservoirs to visualize temporal changes in subsurface fluid patterns. Yet for the most part, these programs remain functionally isolated and rarely *interact* with downstream information flow. As a result, only small portions of the information are passed along to other users.

To understand ways in which GIS can expand a company's ability to convey and analyze information, let us consider the interaction among various personnel and the type of information systems in place.

Information Flow

Business organizations utilize decision-making procedures to reduce uncertainties to improve a company's chance of survival. These organizations work through intelligence-gathering networks that turn raw data into information, information into knowledge, and finally knowledge into decisions (see Fig. 1–1). Geotechnicians, for example, compile raw data sets so that geologists can produce subsurface models. Then engineers drill wells and test promising formations so that economists can evaluate net present value (NPV). Finally, management decides where the company's capital should be directed based on economic comparison of these various competitive prospects.

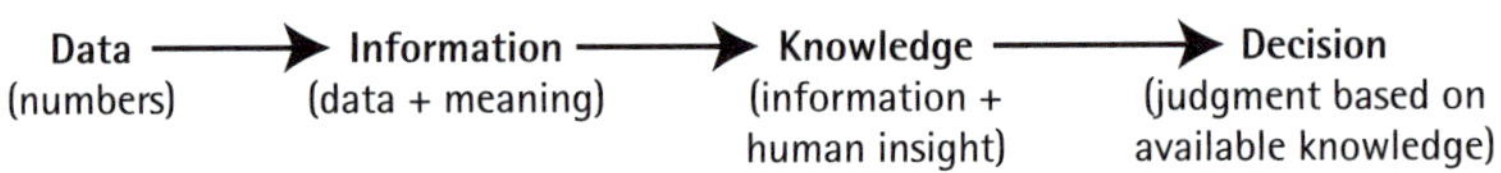

Fig. 1–1 Factors in Quality Decisions

Unfortunately, the serial process of transforming attribute data (aspatial) into information, and the way it is typically localized across the organization, often result in a severe loss of knowledge. As a result, executive managers are not able to evaluate all of the available alternatives in order to make the best decisions. Thus the quality of any decision depends on the quality of the retained information as well as the experience of the technical and management teams.

This is where GIS comes into play. By many estimates, 80–90% of the data used in business and technical work has some type of geographic component.[1] By taking advantage of this relationship, what better way to tie data, information, and knowledge together than through a geographic information system. As such, GIS can help throughout the lifecycle of a project by providing a means to filter information as needed. In some cases, it may even serve as a glue to tie incompatible applications together (see Fig. 1–2).

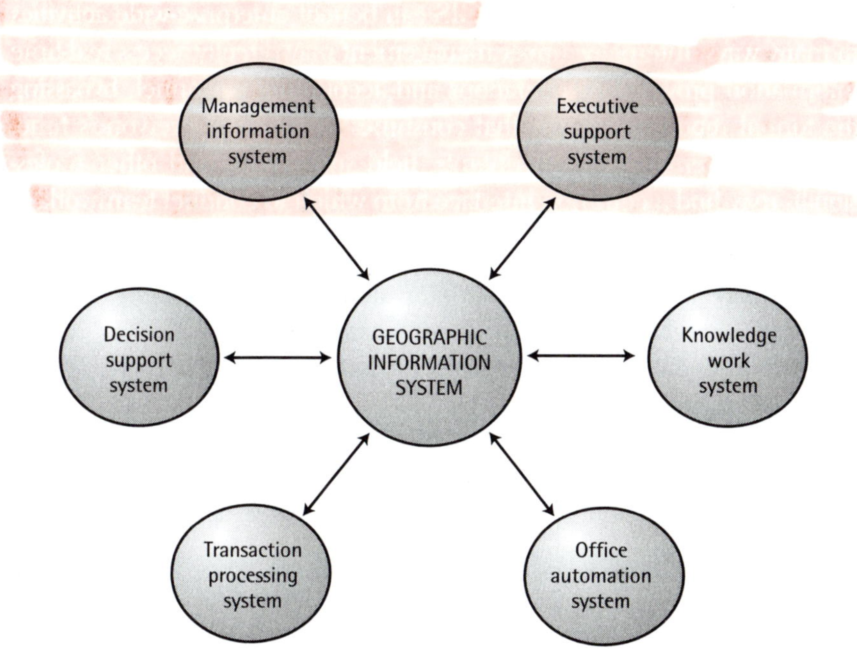

Fig. 1–2 GIS—Central Information Hub

Many readers of this book may question such a concept, but examples of enterprise-wide GIS applications have clearly been shown in the electric, telecommunication, and transportation industries.[2] In such cases, GIS links databases, applications, and users into congruent networks so that work orders can be relayed from the

Seventy-nine percent of data across the business life cycle within Shell E&P has a spatial element.

—GITA 2000 Proceedings

office to the field. Additionally, it has provided new ways for these industries to perform location planning, distribution analysis, and facilities management.[3]

For the oil and gas industry, GIS can benefit enterprise-wide activities in many ways. Internally, upper management may directly access real-time information provided by operations and accounting personnel, bypassing the formal reporting process that consumes so much of everyone's time. Additionally, engineers, accountants, field supervisors, and other professionals may find a common interface from which to conduct teamwork.

Taken from an external perspective, upper management may more easily communicate company activities to shareholders and stakeholders while providing better ways to appraise merger opportunities. It even becomes possible to share information with service and equipment providers.

Definitions

Two things should be apparent: (1) we aren't clear about what GIS can do, and (2) we desperately need to be more clear.

—Joseph K. Berry

To gain a better understanding of GIS, let us begin with a few definitions. As you read along, please note how each definition focuses on a different aspect.

The Association for Geographic Information (AGI) states that GIS is "a *computer system* for capturing, storing, checking, integrating, manipulating, analyzing and displaying data related to positions on the Earth's surface."[4]

The Environmental Systems Research Institute (ESRI), however, says that "a geographic information system (GIS) is an organized collection of computer hardware, software, geographic data, and *personnel* designed to efficiently capture, store, update, manipulate, analyze, and display all forms of geographically referenced information."[5]

DeMers, on the other hand, hints at its role as a *decision-support* tool. "In the broadest possible terms, geographic information systems are tools that allow for the processing of spatial data into information, generally information tied explicitly to, and used to make *decisions* about, some portion of the earth."[6]

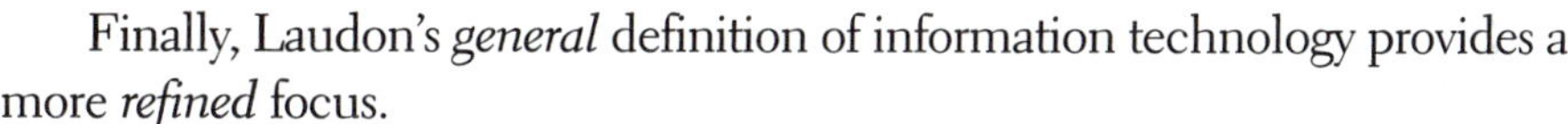

Finally, Laudon's *general* definition of information technology provides a more *refined* focus.

> *An information system can be defined technically as a set of interrelated components that collect (or retrieve), process, store, and distribute information to support decision making and control in an* organization. *In addition to supporting decision making, coordination, and control, information systems may also help managers and workers analyze problems, visualize complex subjects, and create new products.* [7]

As can be seen, definitions vary from source to source, depending on the user, technology system, business or professional use, and the branch of research or industry focus. This issue becomes increasingly complex as one tries to describe or interrelate the numerous sister GIS technologies. These technologies include computer cartography, photogrammetry, automated mapping and facilities management (AM/FM), and spatial decision support systems (SDSS). But for all practical purposes, you the reader should simply focus on the abilities of GIS. Visualize what can be done with this technology, rather than try to work within the guidelines of some restrictive definition.

More than a Map

Think about it. The petroleum industry can and should be examined from a geographic perspective. It is inherently spatial. Each and every oil well, pipeline, tanker, refinery, and gasoline station can be pinned down to a precisely defined position or route that then can be modeled and visually portrayed. To communicate the existence of these assets, we almost always turn to the geographer's best friend, the map. Without maps, a company would very likely lose track of its leases, drill more dry holes than it should, and fail to evaluate changing markets. For our industry, it is an irreplaceable medium, and few industries produce more maps than we do.

GIS allows petroleum enterprises, or functional groups within, to communicate information and make spatial and temporal decisions about assets, activities, and natural resources.

Unfortunately, paper maps serve only as a means to communicate a miniaturized, minimized, and myopic version of the world. Through the physically restrictive nature of this medium, a great deal of aspatial information has to be filtered, displaced, and removed to deliver a useable product. For example, if we tried to show a major oil company's distribution networks on a 36-in. by 36-in. piece of paper, we could do it. But the physical limitations of the paper would force us to leave off many important details needed to describe these assets in full. Within a short period of time, moreover, construction activities, acquisitions, and divestments would make the map out-of-date, requiring time-consuming revisions.

With GIS, however, it becomes possible to tie intrinsically all of the geographic coordinates for each asset to its descriptive "attributes" within a relational or object-based database (see chapter 3 and Fig. 1–3). These attributes are linked to the geographical location of each asset, each uniquely identified by a Cartesian or geographic coordinate (see chapters 4 and 5). As such, we can then query and perform spatial analyses on a virtual model of the asset itself in relation to other facilities and natural resources. Then by keying off the position of the asset(s), we can data mine, or focus our data search, down to the level of a circuit board or scrutinize capital expenditures.

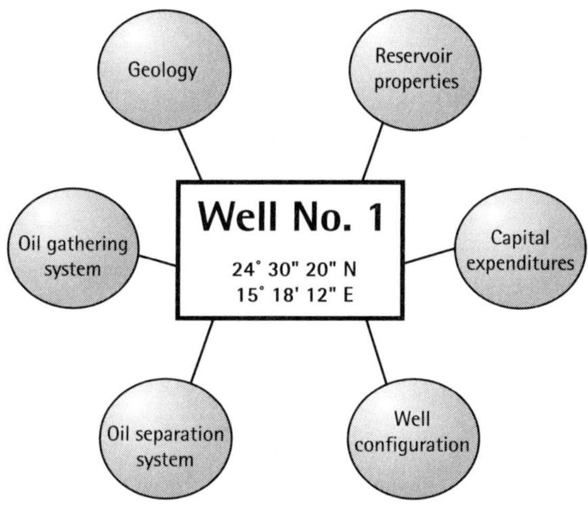

Fig. 1–3 GIS—Linking Attributes to Geographic Position

We may even ask questions of other assets around it, and in relation to it, thereby generating new ways in which to view infrastructure and markets. In turn, an almost never-ending series of interactive maps can be generated from the database through mathematical manipulation of computer-generated layers, either through graphical means or by tabular inputs.

With GIS, we can answer the following questions:

- Where should we drill?
- Where should we lay a pipeline?
- Is it cheaper to transport our crude via pipeline, railroad, or tanker?
- Which world crude oil markets provide the highest price differential for a futures contract?
- Which refineries can handle heavy oils with sulfur or light and sweet?
- How should we tailor our reformulated gasoline runs in accordance with seasonal market demand or regulations?
- Who can we sell our petroleum products to?
- What is the cheapest route to get products to market?
- Where should we build new gasoline stations?
- What corporate assets should we keep or divest during a merger to take advantage of operational synergies or comply with anti-monopoly rulings?

With a geospatially anchored database, we can locate features on and within the Earth, search for geographical patterns, determine optimum locations and corridors, analyze changes over time, and even decipher cause-and-effect relationships.

Workflow and Information

At a fundamental level, the essential contribution of information technology is the expansion of knowledge and its obverse, the reduction in uncertainty. Before this quantum jump in information availability.... Businesses had limited and lagging knowledge of customers' needs and of the location of inventories and materials flowing through complex production systems.[8]

Alan Greenspan,
Chairman, U.S. Federal Reserve, 2000

To see why Mr. Greenspan's insights are so intuitive, let's examine the organizational structure of a contemporary oil company and relate it to the type of information network typically in place.

First, it can be proposed that company workflows are a direct reflection of the structure of its information network. For example, if the network consists of a series of secularized information systems (tough to call this a network), then information and workflow tend to be serial in nature (see Fig. 1–4, top). Accordingly, the process of turning data into knowledge becomes a matter of throwing information "over the wall," from one functional group to the next. In this manner, the company ends up with discrete pockets of knowledge that lead to imperfect decision making.[9]

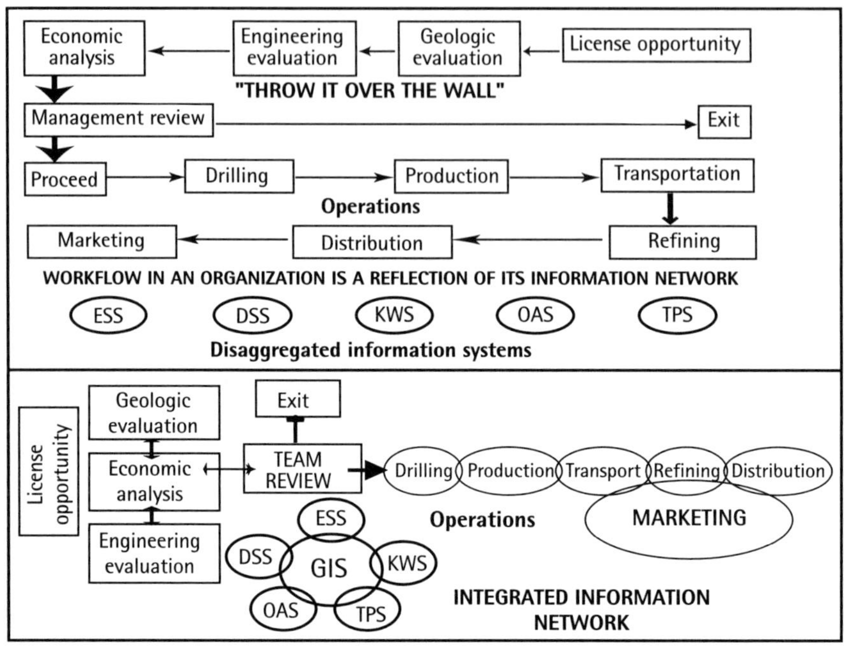

DSS=Decision support system; ESS=Executive support system; KWS=Knowledge support system; TPS=Transaction processing system; OAS=Office automation system; GIS= Geographic Information System

Fig. 1–4 Workflow and Organizational Structures

If concurrent business practices are put into place, however, where team members across the organization communicate and work together more closely, then a better approach can be applied. This is shown in Figure 1–4, in the bottom diagram. Please refer to the following acronyms, as adapted after Laudon:

- DSS—Decision support system
- ESS—Executive support system
- KWS—Knowledge support system
- TPS—Transaction processing system
- OAS—Office automation system
- GIS—Geographic information system[10]

In essence, disaggregated systems produce serial workflow, whereas integrated systems aid concurrent engineering. Take for example the way in which upstream companies traditionally evaluate and develop a prospective oilfield. In most cases, geologists, engineers, marketers, and economists remain at arm's length from one another, each working within the traditional roles passed down during the past 40 years (see Fig. 1–5, top).

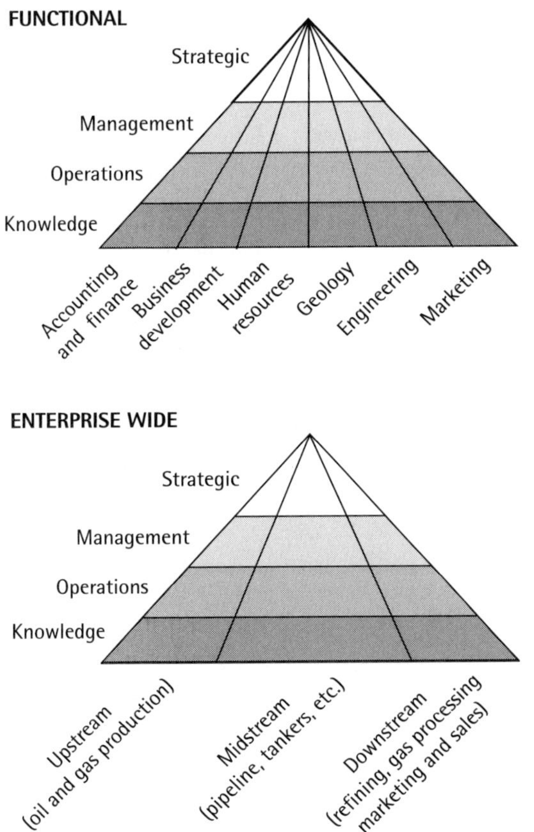

Fig. 1–5 Specific and Enterprise-Wide GIS Applications

Without question, geologists at times work closely with engineers, but only grudgingly, and usually with limited teamwork or information sharing. And as the downstream workflow widens between one professional and the next, the level of interaction decreases.

Geologists and marketers, for example, spend little time evaluating prospects together, although the success of the project depends both upon natural resource distribution and the ability to move the product to a sales destination. Such nonrelationships in turn affect the culture of the organization, resulting in role clashes and individualized goals that often differ from the company's mission. Consequently, such professional isolation severely hampers teamwork.

But if the company has developed an interoperable information system data based on geographic relations, a technological platform for parallel and interactive workflow can be put into place across the organization. (Interoperability, which is a measure of the ability of a binary system to meaningfully exchange and use data, refers to the quality of the connection between multiple, heterogeneous systems, subsystems, or components.[11]) This provides the company with an ability to concurrently develop and retain all available information (see Fig. 1–5, bottom). Although such parallel processing requires a tremendous amount of up-front work, the need to reengineer down the road is mitigated.[12] As all engineers know, it is easier to fix a problem at project start, when the design is simple, than later on, when numerous systems have become inextricably intertwined.

Furthermore, this ability, through geographic-based project analysis and cross-company communications, may allow geologists, engineers, and marketers to work more closely together, beginning at project inception. Although such a concept is obvious to most, many organizations continue to treat geology, technology, engineering, and marketing as independent endeavors, rather than as complementary functions.

In summary, a major inhibiting (or promoting) factor related to project life cycle, workflow, and profitability is the type of information network in place. If there are several systems, each working independently, then the individuals working with these systems will also be isolated. But if everyone utilizes the same network, and the human part of the organizational structure takes advantage of the system, then everyone can feed off and contribute to the same databank. As a result, the company can obtain a competitive advantage without spending a dime on manufacturing competencies.

Real-World Examples

To provide the reader with a feel for GIS in everyday applications, the following section diverges from the main focus of this book. Descriptions are given here of a few GIS applications that affect our daily lives, rather than focusing on petroleum applications. By taking this side trip, however, your understanding of GIS should begin to grow.

Finding your way

By now, most Internet users have taken advantage of the many driving instruction utilities provided by various mapping sites and search engines on the Web (i.e., www.mapquest.com, www.yahoo.com). Behind the scenes, driving instructions provided on the Internet are actually a rudimentary form of GIS. In essence, this technology works in the background by using a geocoded database that determines geographic position through address and zip-code based queries. The user simply requests directions by filling out text boxes.

Once the query is complete, the GIS calculates the shortest path to the destination along with alternatives provided in text and map form. Generally, the instructions dictate where to make a turn indicated by distance and drive time from intersection to intersection. From a graphical perspective, a scalable map provides anywhere from national highway to street level detail.

Luxury automobiles and trucking companies also commonly utilize GPS-enabled navigation systems that track the vehicle's position in real time. For the former, a dash-mounted, street-level display shows the requested route through keyboard or voice-based queries. In some systems, verbal directions also prompt the driver when to make the turn so that the driver is not distracted by the console-based map. If the driver misses the exit, some units will even recalculate the route automatically.

For the trucking industry, the technology involves wireless communication systems, GPS, and GIS so that a firm can keep track of its fleet, optimize schedules, and reduce fuel costs. Firms also can improve safety by ensuring drivers obey the speed limit and pull over when it is time to sleep. As these two examples show, GIS provides both individual and enterprise-wide solutions.

Real estate

Another example of GIS has been provided by the real estate industry. The search for new or used homes, as most homeowners know, can be a frustrating experience involving time-consuming onsite visits. Fortunately, some Internet-based GIS tools now help to narrow the search.

One utility, for example, uses map interfaces at a nationwide level that allow you to graphically "mine" down to the street level (www.realestate.com). Searches are conducted by querying homes within a certain zip code or range of zip codes. Next, you choose the value range of the house, the number of bedrooms, and so forth. Afterwards, you are left with a list of potential homes with links to floor plans and photos for each house. Then you can make a decision on whether to physically inspect the house or not.

If the real-estate firm is truly GIS savvy, it is even possible to search more specifically. Searches can include homes within a certain distance from a school (see chapter 3 discussion on buffers), areas that have a low tax rate, or even can determine which areas are prone to flooding.

U.S. elections

An excellent example of GIS analysis can be shown from the 2000 presidential contest between candidates George W. Bush and Al Gore. The precinct map shown in Figure 1–6 (Plate 1–1)[13] shows how the dispute between the Electoral College and popular vote can be analyzed from a geographical point of view. As reported by the media, Gore won the popular vote. U.S. elections, however, are based on the Electoral College, and if one tallies the winner for each precinct across the entire country in a map view, a different perspective can be appreciated. (Color plates for each chapter are located in the center of the book.)

Unfortunately, such an analysis also shows how geographic information can be used to portray a particular point of view for political gain. Therefore, a great deal of thought and care must be considered before using GIS as an objective tool for communication and analysis.

On an equally important level, GIS is now being used by the government to help create senate, house, education, and judicial districts based on TIGER, U.S. Census, and election data.

(TIGER stands for Topologically Integrated Geographic Encoding and Referencing, which provides geographic data and statistics on population, age groups, minority groups, etc. from the U.S. Census Bureau.)

The Texas Legislative Council (TLC), for example, has developed a GIS program that consists of three applications. These are: Redistricting Application (Red Appl), Boundary Definition System (BDS), and a Spatial Integrated Cartographic Environment (SPICE).[14] BDS provides a user interface for the TLC to create and update cartographic databases primarily for precinct coverage, voter tabulation districts, and school district boundaries.

Thus, GIS now affects us on a much grander scale than previously imagined. Each district, for example, must:

1. Follow a block boundary line
2. Be contiguous
3. Have a population that is close to the population of all other districts in the plan
4. Not have a minority population that will have the effect of discriminating against the minority group(s)

Obviously, GIS provides a tool for reapportionment and affects such things as school funding and the political makeup of the nation. Some of the map outputs include school districts, school funding, census geography, election results shown through thematic shading, and a redistricting plan portrayed by a population theme.

On a federal level, the American Community Survey (ACS) has been established to implement a continuous collection and measurement system that would produce annual and multiyear estimates of population characteristics. In 2003, the U.S. Census Bureau planned to implement the ACS in every U.S. county and Native American area. The program will be used to identify and address community issues. Such issues include health care, public safety, education, economic development, elderly needs, child welfare, rural issues, and transportation (www.esri.com/new/arcnews/winter0102articles/uscensus-bureau.html).

Oilfield scouting with GIS

Now let's take a look at a GIS application used by coalbed methane (CBM) workers in northern Wyoming. In the years 1997–2000, CBM activity in the "Cowboy State" increased from 229 to 1374 producing wells.[15] Correspondingly, production increased eightfold to 225 MMcfd.

Yet behind the scenes of this drilling boom, an old breed of driller has begun using advanced Internet, GIS, and satellite navigation technologies (GPS) in some very innovative ways. "Workers are using these technologies to watch what each other is up to," said Bruce Lambertson of Mountain Sports, Casper. "In essence, this has brought the art of scouting to a new level."[16]

Rick Marvel, engineering manager for the Wyoming Oil & Gas Conservation Commission (WOGCC), says the state website (http://wogcc.state.wy.us) recorded 1000 hits/day between 8 A.M. and 5 P.M. in 2001. On this site, WOGCC provides well coordinates, production data, permit approvals, geological markers, and so on, made available to anyone who wishes to use them. Since operators can track each other through the website, looking for tops, casing points, and perforation intervals, the learning curve has been accelerated.

Marvel says the commission provides the means to download or hyperlink information to a variety of software programs. For example, "Production data and coordinates can be easily imported into Geographix (Landmark) and Excel."[17] Additionally, a WOGCC link to a topographic mapping package (www.alltopo.com) directly places well spots on $7^1/_2$-min. topographic maps. This includes right-click hyperlinks back to the website for additional information such as production. Finally, there is a GIS front end located at the University of Wyoming's Internet map server (http://www.sdvc.uwyo.edu/). It allows users to build custom-designed base maps without installing a GIS on each personal computer.

GIS, GPS, and Internet technologies are also providing new tools for surface planning and facility management. "By downloading GPS waypoints into (a GIS), I can go out in the field and spot new roads and pipeline routes at 50 mph," said Larry Brown, owner of H&B Petroleum Consultants. "While it doesn't replace the surveyor, it's great for planning."[18]

In the future, these tools will become more important as operators strive to overcome water problems associated with CBM. "The reason why the economics looked so attractive (in the beginning) was because there was no water-handling expense," said Darrick Stallings of Yates Petroleum Corp. "We could simply pump the fresh water to the nearest drainage."[19]

Unfortunately, there is little need for drinking quality water in a state of 450,000 people. He said that with the unexpected water production volumes now being produced, "the costs paid to owners for storage and erosion damage can drive total well cost upwards of $100,000."

Tools of the Trade

Throughout this chapter, we touched on several subjects ranging from various definitions of GIS to enterprise-wide applications. Fortunately, the petroleum industry has been blessed with a variety of tools to collect, process, and analyze geographic data. The remainder of this book is designed to help you understand the basic fundamentals of these technologies, allowing you to take advantage of GIS's abilities.

In chapter 2, three oilfield examples will be described to show the utility of GIS in the upstream, midstream, and downstream sectors. In chapter 3, "Spatial Data and Modeling," the differences between point, line, area, network, and surface entities will be examined. The subject matter then turns to vector and raster data and the strengths and weaknesses of each. Finally, the geometric relationships between objects in space based upon adjacency, containment, and connectivity—otherwise known as topology—will be described.

In chapter 4, "Geodesy," one of the most misunderstood topics in the petroleum industry will be discussed in depth. Without sound knowledge of this subject, any use of GIS can be fraught with problems if the user accidentally commingles data from various sources. Chapter 5 carries on by walking you through the subject of projections and grids.

Chapter 6 then turns to the subject of the GPS, a satellite-based technology that benefits petroleum work in many ways. In chapter 7, remote sensing technologies as used in the oil industry will then be described.

Finally chapter 8, "The Art of Presentation," explains the complexity of map design. It shows how the map projection, frame of reference, level of generalization, features, scale, and style can be used to produce a powerful presentation tool. Conversely, it describes how incorrect representation can lead to poor decision making.

References

1. Moloney, T., A.C. Lea, and Kowalchuk. *Manufacturing and Packaged Goods, In Profiting from a Geographical Information System*. Fort Collins, CO: GIS World Book. pp. 105–129.
2. von Meyer, N.R. and R.S. Oppman. 1999. *Enterprise GIS*. Park Ridge, IL: Urban and Regional Information System Association (URISA). pp. 93–98.
3. *2000 Geospatial Technology Report: A Survey of Organizations Implementing Geospatial Information Technologies*. 2000. Aurora, CO: Information Technology Advisory Group, Geospatial Information & Technology Association (GITA). pp. 17, 33, and 37.
4. Association for Geographic Information. Website. (www.agi.org.uk/).
5. Environmental Systems Research Institute. Website. (www.esri.com; glossary at http://www.esri.com/library/glossary/e_h.html#GIS).
6. DeMers, M.N. 1997. *Fundamentals of Geographic Information Systems*. New York: Wiley and Sons. p. 7.
7. Laudon, K.C. and J.P. Laudon. 1996. *Management Information Systems—Organization and Technology*. 4th ed. Saddle River, NJ: Prentice-Hall. p. 9.
8. Greenspan, A. 2000. "The Revolution in Information Technology." Remarks before the Boston College Conference on the New Economy. Boston, MA (March 6).
9. Roussel, P.S., K.N. Saad, and T.J. Erickson (Arthur D. Little). 1991. *Third Generation R&D—Managing the Link to Corporate Strategy*. Boston:Harvard Business School. p. 5.
10. Laudon and Laudon, p. 19.
11. von Meyer, p. 13.
12. Wilemon, M. and D. Wilemon. 1991. "Determinants of Cross-Functional Cooperation in Technology-Based Organizations." *Journal of Engineering and Technology Management*, Elsevier. 7:19. pp. 229–250.
13. *USA Today*. 2000. (November 9).
14. Clark, et al. Website. (www/esri.com/library/userconf/proc96).
15. Gaddy, D.E. 2000. "Journally Speaking." *Oil & Gas Journal* 98:19 (May 8) p.15.
16. Lambertson, B. Oral communication.
17. Marvel, R. Oral communication.
18. Brown, L. Oral communication.
19. Stallings, D. Oral communication.

2

Oilfield Examples

Oil is found in the mind,
not in the ground.

—Famous oilpatch quote

In the petroleum industry, an infinite number of decisions and operations may be supported through some sort of geographical assessment. Upstream, a land department may use GIS and computerized cartography to keep the company appraised of lease commitments. Midstream, a pipeline engineer may use GIS, GPS, and inertial navigation systems (INS) to map material anomalies along a pipeline.[1] And downstream, a marketing manager may use GIS with network nodal analysis to choose optional transportation routes.

With these kinds of analytical tools, it is important to ascertain the most appropriate way in which to solve a problem. This in turn requires a different thought process that focuses on dimensional relationships rather than textual.

In all cases, a GIS analysis must begin in the form of a question: Where did we drill our best wells last year? Where should we construct a pipeline? If company A merges with company B, what ideal combination of assets will provide new synergies?

Depending on the user, once the question has been formulated, the means of acquiring the answer will depend on the quantity and quality of accessible data that must be gathered. It also will depend on the hardware and software systems needed to process the data and the GIS technique used to answer the inquiry.

For example, the chief operating officer may require a few digitized maps, an off-the-shelf GIS program, and a stand-alone computer to make a generalized map of the company's operations. On the other hand, the needs of an asset team will differ. Instead they may need access to company-wide engineering and geologic data, a relational database, a Unix-based workstation, and a proprietary GIS to develop a shared Earth model. Obviously, the quantity of information demanded by the asset team, and the technologies required to process and analyze the data, will be more complex. Yet both users have one thing in common—a need to spatially visualize and analyze data for improved decision making.

Other important considerations for GIS development include the scope and global activities of the business. If the company is a small upstream oil producer operating in Wyoming, for example, a basic GIS program installed on a few select computers may satisfy the company's needs. On the other hand, a vertically integrated company with international operations in Asia may utilize an Internet-based mapping system linked to an object-oriented database. Again, the data resolution and the complexity of technologies may differ. Yet the ultimate goal of producing a faster, cheaper, and more accurate answer remains integral to the user's objective.

Generally speaking, GIS can be broken down into five major activities. In its simplest function, GIS can present data in map form to communicate information. Second, GIS can organize geographic information in map, chart, or table form to visualize spatial patterns in order to stimulate visual thinking.

Third, GIS can query geographic points of interest and associated attributes to answer the question, "What and where?" Fourth, it can provide new information by building geographic themes from older layers.

Finally, GIS can track patterns in space and time to aid in the function of analysis, decision making, and workflow. In the following examples, all of these elements are described.

Oil Industry Examples

> *How does a business know the possible applications of GIS if it is not aware of its capabilities?*
>
> —David J. Grimshaw

Numerous times the author has heard criticism of the petroleum industry for its ignorance and "ignore"-ance of what is apparently an underutilized technology. Yet our business has a certain familiarity with spatial information and information technology that goes beyond the technical capabilities of most industries. We should be proud of, and more importantly, take advantage of, this capability.

Geophysicists, for instance, commonly work with 3-D seismic surveys that contain more than a terabyte of data. This volume of data requires tremendous processing power and advanced software programs to filter out noise to produce discernible cross sections. Drillers, moreover, apply differential GPS to maintain station keeping in deepwater floating operations (see chapter 6). One can also visit a U.S. Gulf Coast refinery or Canadian heavy oil facility to see AM/FM and supervisory control and data acquisition (SCADA) technologies at work. As such, it can be said that petroleum high-technology applications do not lag other industrial applications, despite many comments to the contrary. Rather, they take a leading role in the development of information and spatial technologies.

The problem then clearly is not related to the singular use of telecommunication technologies, remote-sensing techniques, computers, or software applications. Instead, the challenge arises in the proper mixing of these technologies to implement a new paradigm. Building an effective GIS should not be treated as an experiment where users try to "cobble" together software or intertwine telecommunication systems with databases. Instead, it is one of trying out new thought processes to solve a particular problem from a geographic point of view through system integration. The petroleum industry has the means to solve its technical goals. Thus it is only a matter of putting the technologies to work in a cohesive manner.

To gain a better understanding of GIS's underlying potential, the remainder of this chapter is dedicated to GIS oilfield examples. The three topics—wellsite selection procedure, pipeline corridor selection process, and market analysis—are by no means exhaustive. Rather they are intended as a source of stimulation for the reader.

Wellsite Selection

A routine problem for exploration and development personnel involves wellsite selection. This type of work involves regulatory, ecological, geological, engineering, and technical issues that traditionally have been solved by overlaying successive maps on a light table.

In the San Juan Basin of New Mexico, for example, operators deal with:

- Highly irregular terrain such as canyons and arroyos (see Fig. 2–1, Fresno Canyon Topo Map, Source iGage Inc.)
- Native American archaeological sites
- Different spacing requirements dependent on production fluid type
- State regulations that limit available drilling acreage
- Infrastructure avoidance issues (i.e., pipelines, telephone lines, and utility lines)

- Infrastructure tie-in issues (i.e., pipelines, gathering stations, compressors, pumps, and separation facilities)

In Figure 2–1, the topographic map may appear to be a simple scan of a USGS survey. Additional layers such as well spots, GPS routes, and other features. However, additional layers can be superimposed on top of the map, providing a tool for GIS analysis. This can be accomplished in a special raster-to-vector program.

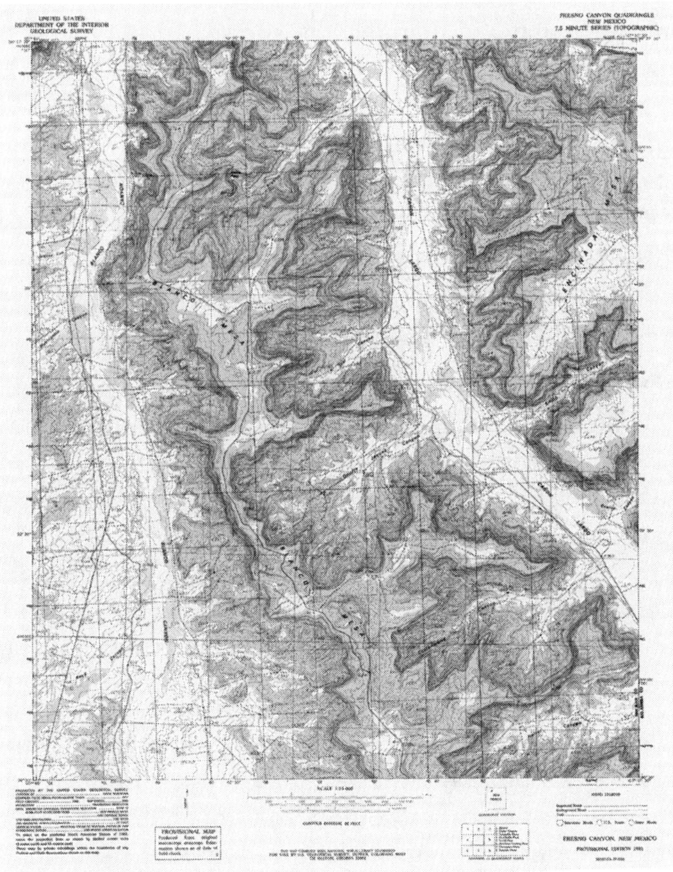

Fig. 2–1 Fresno Canyon Topographic Map [source: USGS and iGage (www.igage.com)]

In New Mexico, some companies select locations by plotting drilling windows onto topographic maps. Commonly, the geologist, aided by a geotechnician or draftsperson, produces a wellbase map on a predefined scale. This map shows existing surface and bottomhole locations symbolized by producing formation and fluid type. Next, various thematic maps consisting of topography, aerial photos, facilities, archaeology, and geology are either drafted, ordered, produced, or plotted on the same scale. In many cases, these maps are already aggregated.

Then in a systematic manner, the site selection process begins by overlaying these maps to identify the most promising sites. By laying the well-base map on top of the topographic map, for example, high-relief slopes can be visually eliminated. In the same manner, aerial photos can be used to identify roads, facilities, and other features to choose access routes or select sites for infrastructure development. This process continues by overlaying archaeology maps, inscribed with radii of avoidances, on top of the evolving location map. Then geologic and production maps are amalgamated to identify production windows.

Finally, the geologist considers the spatial problem of well spacing. The state of New Mexico, for example, restricts gas wells in the Pictured Cliffs to 160 acres (some exceptions are granted). On the other hand, oil pools may be developed only on 40-acre spacing. To further complicate matters, gas wells in San Juan and Rio Arriba Counties cannot be located closer than 790 ft to any outer boundary of the tract. Nor can they be closer than 130 ft to any quarter-quarter section line or subdivision inner boundary. Thus, these tracts are confined to about 9% of the total square feet in the section.

In light of all the geographic information that must be commingled, the manual overlay process turns out to be a time-consuming process. It also is prone to mechanical error induced through dissimilar map scales, copy machine distortion, map stretching, and edge matching (see Box 2–1).[2, 3, 4, 5] Furthermore, human errors occur throughout the process as one map is "eyed" against the other and as all the regulatory restriction creates confusion.

Box 2–1

Edge Matching and Rubber Sheeting

When dealing with a study area that extends across two or more map sheets, mismatches between adjacent map sheets will take place. A river, for example, may not line up, or satellite images taken at different times of the day, and under different weather conditions, may produce artificial differences (Heywood, p. 96).

To overcome this problem, the digital map must be edited in the boundary zones to ensure that there are no breaks in features (Jones, p. 89). One process, called *edge matching*, compares and adjusts features along adjacent edges to ensure they agree in both positional and attribute terms (McDonnell and Kemp, p. 33). In such cases, each sheet should be edited, reprojected into the same cartographic system, and then digitized separately so they can be joined seamlessly.

Certain data sources may also occur due to internal distortions within the individual map sheets (Heywood, p. 97). This is particularly troublesome for data derived from aerial photography as caused by asymmetric aircraft movements and distortion caused by the camera lens.

These inaccuracies, which may remain even after transformation and reprojection, can be rectified through a process known as *rubber sheeting*. Rubber sheeting involves stretching fixed points towards established control points much like an elastic sheet. These control points should be easily identified on the ground or in the image (i.e., outcrops, distinctive buildings, mountain peaks, etc.) as geographically determined from field surveys or by using GPS.

Obviously, GIS provides an alternative approach. One useful technique includes the algebraic selection of well locations with the assistance of computer models like structured query language (SQL).

Binary modeling

Landslide susceptibility models, which utilize themes familiar to petroleum geologists, have long used digital modeling to predict the areas most prone to avalanches. Joseph K. Berry of the University of Denver geography department clearly describes a raster-based binary model that replaces the manual overlay process.[6] In the simplest model, logic is employed to describe one of two states: *Yes* and *No*. Three layers consisting of elevation, soil, and vegetation cover are first digitally gathered, processed, and geospatially related to one another. Each theme is then qualified to meet specific criteria. For example, after calculating the first derivative of slope from a digitized topographic map, all slopes greater than 30% can be identified. Following this methodology, soils and vegetation types that promote excessive mass wasting can be rated as well.

Representing simple divisions, binary codes using 0 or 1 are then inserted into the raster cells of all three layers. In such a manner, those conditions that potentially promote landslides can be digitally and spatially qualified (see Fig. 2–2).

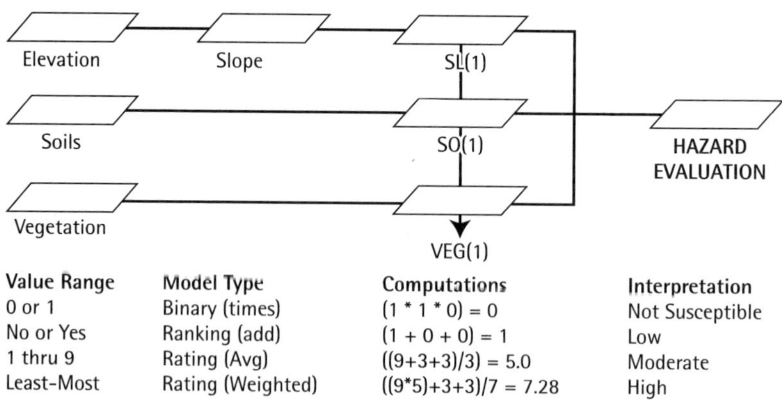

Value Range	Model Type	Computations	Interpretation
0 or 1	Binary (times)	(1 * 1 * 0) = 0	Not Susceptible
No or Yes	Ranking (add)	(1 + 0 + 0) = 1	Low
1 thru 9	Rating (Avg)	((9+3+3)/3) = 5.0	Moderate
Least-Most	Rating (Weighted)	((9*5)+3+3)/7 = 7.28	High

Fig. 2–2 Landslide Potential (source: adapted from Joseph K. Berry[7])

Then it becomes a mathematical game. For example, underlying cells of the elevation and soil layers for a given area may be identified by 1 (susceptible), but the vegetation layer positioned in the same exact location is 0 (not susceptible). Consequently, this particular piece of real estate may be evaluated as nonsusceptible to landslides through multiplicative operands, i.e., 1 x 1 x 0 = 0.

The beauty and simplicity of this model is that it can be expanded to include ordinal and ratio analyses that rank and rate landslide susceptibility, further explained in chapter 3.

For an oil and gas company, a similar process can be applied in the San Juan Basin as well:

- Well coordinates can be acquired from surveys or GPS receivers.
- Geologic and production data can be geographically tied to well coordinates and subsurface surveys and then manually entered into a spreadsheet for import into a relational database.
- Topography and aerial photographs can be ordered from the United States Geological Survey (USGS) as digital elevation files (DEFs) or digital orthophoto quads.
- Archeological maps can be digitized from the paper maps developed during environmental impact studies.
- Regulatory restrictions can be written as scripts using Boolean logic that defines spatial relationships.

Once this information becomes resident in the GIS database, primary data layers can then be merged, joined, buffered, or dissolved to produce entirely new layers.

Corridor Selection

For corridor selection, GIS has been successfully applied in the determination of pipeline routes. Although it cannot replace actual field surveying, preliminary GIS planning can help identify immediate problem areas, streamline field activities, and reduce development costs. Furthermore, GIS has also been used as a link between nonstandardized software programs and data formats.

Preliminary planning

Dominion E&P Inc. and Geoscience Earth and Marine Services Inc. (GEMS) used GIS as a preliminary pipeline planning tool for the Devil's Tower prospect in the Gulf of Mexico [see Fig. 2–3 (Plate 2–1)]. According to John Brand, a geologist with GEMS, "The old method of planning would have been to draw a straight line from the prospect to a possible host platform, followed by a field survey of the route to determine suitability." Using GIS, however, he immediately identified a number of features that made the straight route undesirable. "These features included man-made structures and (natural) topography that would have made this route technically unfeasible."[8] In other words, a straight-line route would have crossed pipelines, wells, and an explosive ordnance dumping area.

Software issues

The data needed to evaluate these routes came from a number of sources: government agencies, historical archives, and proprietary information. Each particular data set required a particular software program. Seismic Micro-Technology's Kingdom Suite, for example, contributed digitized geologic interpretations to GIS via AutoCAD DXF files while providing ASCII seismic attribute data to Golden Software's mapping program, Surfer.

Surfer was then used to generate grids such as bathymetry, seafloor amplitude, and positive or negative seafloor slopes, depending on the seismic data. These grids were then imported into the GIS program MapInfo for final interpretation.

"If the software cannot directly provide the desired data format, odds are another one can," said Brand.[10] Subsequently, a number of scenarios were designed to avoid the ordnance dumpsites and minimize pipeline crossovers. As a cost alternative, optional routes were designed to pass through the dumpsite while paying heed to steep or rocky seafloor.

"Each preliminary route is dynamic and can be updated 'on the fly' so that attributes can be used to assess each route's cost-effectiveness."[11] The end result, however, is not only a controlled data stream, but a system that allows petroleum professionals from various disciplines to work with one another. This in turn leads to more efficient planning and workflow.

Marketing

Business has traditionally focused on questions relating to what (to produce), how (materials and technology), and why (strategy), neglecting the question of where (to locate and to find customers).

—David J. Grimshaw

Perhaps the greatest potential for applying GIS in the petroleum industry lies in marketing. GIS can be used to identify, evaluate, and target markets for sales of crude oil, gas, and products. In 1999, for example, Arco (later acquired by BP plc) utilized a GIS approach to analyze the marketing potential of a proposed Chevron Inc. merger in the West Texas area. The main focus was to evaluate transportation options to determine the best way to move crude production to U.S. refineries, explained Curt Shill, a former marketing manager with Arco. He told attendees at ESRI's Petroleum User's Group, "There are three important conditions to consider: the location of your crude, the quality of your crude, and the time of delivery."[12]

To take a look at the potential synergies and divestment strategies between Arco and Chevron, Shill utilized several databases. These included U.S. crude supply pipelines (MapSearch, PennWell), the Annual Refinery Survey from the *Oil & Gas Journal*, Tobin's land grid, and internal information from Arco and Chevron. From this information, he evaluated various options:

1. Utilize existing assets to get crude to Midland Trading Center or Shell Odessa's hypersweet crude refinery
2. Utilize either an underutilized Mobil pipeline or Texas-New Mexico idle sour pipeline to reach regional refineries
3. Evaluate the economic prospect of local trucking

Although the content of this information was not provided for public use, Shill's analysis provides a hint as to the ability of GIS in marketing. Shill also says that GIS can provide insights into the changes over time in:

- Refinery shutdowns
- Refinery capacity
- Pipeline
- Leases marketed
- Transportation costs from lease to market centers

All of these depend on the availability and quality of data. As the following example illustrates, users can employ individual, public, and private datasets to make a complex problem easier to understand and solve.

Russian marketing

GIS can be used to expand the utility of spatial visualization and network relationships in marketing analysis. For example, the author used GIS to sift through the various transportation and sales alternatives associated with selling Russian crude. As it stands, the former Soviet Union (FSU) has a very complex distribution system that simply cannot be described from a textual point of view. Complicated relationships exist among oil fields, storage facilities, processing systems, metering points, pipelines, refineries, and seaports. Plainly these relationships are just too difficult to ascertain effectively and quickly without GIS. Moreover, the need to sell to three markets instead of one adds another dimension of complexity.

In Russia, producers and traders must optimize their sales portfolios with the following:

- Domestic markets. Sales to local refineries within the borders of Russia
- Near abroad markets. Sales to refineries and offtakers in the FSU countries outside Russia (i.e., Belarus, Ukraine)
- Far abroad markets. Sales to refineries and offtakers in continental Europe or seaborne destinations outside Russia and the FSU

In 2001, roughly 40% of a typical upstream oil company's crude oil sales took place on the domestic market, 15% on the near abroad market, and 45% on the far abroad market. (This excluded product sales using domestic crude feedstock.) During 2002, however, the ratio of far abroad to domestic sales began to drop as incremental production came on-line from the large capital investments made by the Russian majors. In 1996, for example, Russia produced 6.1 million BOPD, rising to more than 8.0 million BOPD in 2003.

This created a more intense situation as competing producers strove to sell every drop possible to Western markets, as these afforded higher netbacks. Unfortunately, several factors that haunted producers throughout the 1990s remained in play into the early part of the 21st century. These negative factors included pipeline constraints, a shortage of all-weather ports, and compulsory supplies to industrial and agricultural activities. Other negative factors included modest domestic consumption and oversupply to local refineries. As such, the government continued the quota system by allocating far abroad export volumes to each company based upon difficult-to-substantiate reserve and production capabilities. This had the effect of limiting their freedom to flexibly sell crude on the open market.

In principal, every producer was entitled to sell 30% of its crude production to the world markets via the main export schedule. Once each company's export schedule was announced, the Russian government, through the Central Dispatch Unit and Transneft, then determined whether supplemental shipments could be sold to other foreign destinations. This determination obviously tried to consider supply-and-demand issues. These issues involved pipeline capacity, inventory levels, throughput volumes, domestic consumption, upgrades, harvest season, weather events, and intentional squeezes. Based on these calculations, supplemental shipments to the far and near abroad were granted ad hoc to applicants, taking into account social obligations and political arrangements.

Once Transneft published these schedules, each company then focused on selling its remaining production to the domestic market. This was accomplished by finding those refineries that would provide the highest netback, with two provisos. (Netback from a producer's standpoint equals sales price minus transport costs, blending costs, taxes, and commissions.) First, if the company had to fulfill its own corporate demand for refined product, the feedstock was passed directly from the producing field to the company refinery. Second, for those companies with no local refineries, or those with excess inventory, these volumes would be sold on the small, yet growing spot market.

Obviously, understanding the *aspatial* issues of politics requires an in-depth understanding of the players in the Russian oil business. But just as important, it is necessary to *spatially* understand geographic relations among the upstream, midstream, and downstream sectors to make sound decisions. Consequently, the remainder of this chapter will concentrate on domestic oil sales opportunities and far abroad transportation costs through the eyes of GIS.

Domestic markets

Let us assume the role of an oil company operating in the Volga-Urals region, working near the Romashkinskoye oil field [see Fig. 2–4 (Plate 2–2)].

The utility of GIS becomes quite apparent in our analysis of the oil company's various markets. Features like pipelines can be denoted with symbols that indicate crude oil flow, diameters, and capacity. Then by using network analysis programs, nodes and distance vectors can be used to determine the shortest or cheapest path to each destination.

In the Volga-Urals oil province, particularly, market changes have begun to take place much faster than in West Siberia. This provides Volga-Urals companies certain advantages over competing companies operating east of the Urals. There are four basic reasons for this.

First, the Volga-Urals oil province has an established network of buyers and sellers that conduct business through a growing number of producing entities, traders, and refineries. Thus, independent and integrated oil companies alike have numerous choices from which to sell crude oil to refineries in the European sector of Russia, the near abroad, and Eastern Europe. These sales take place both as spot and term contracts typically concluded at some fixed price.

In comparison, the regional monopolization of West Siberia by large vertical companies has inherently limited market development. This is because most oil moves directly from the wellhead straight to the refinery without leaving the jurisdiction of the oil company. Thus, location in Russia plays a role in market access.

Second, Volga-Urals producers are located closer to major consumption areas than West Siberian producers. Thus, transportation costs are cheaper, which is predominantly a function of distance. For producers in Tatarstan, for example, it is only 650 mi to Moscow from Almetyevsk. On the other hand, it is 1500 mi from Nefteyugansk, Siberia.

Third, the direction of pipeline flow across the Transneft pipeline system dictates whether a seller can reach a buyer or not. Generally speaking, oil moves east to west (towards Europe), northeast to southwest (towards the Black Sea), and southeast to northwest (towards the Baltic Sea). Therefore, unless two companies utilize a physical swap, the problem of finding sales destinations within Russia depends on downstream access to storage facilities, ports, and refineries. In this sense, producers operating in the Volga-Urals area have access to more trunkline junctions than those operating in West Siberia.

Fourth, crude quality often determines where a particular crude type will be sent. To maintain some level of purity, Transneft may force heavy, sour oil producers to pump crude on dedicated pipelines to Kremenchug, Ukraine; Ufa, Bashkortostan; or Ventspils, Latvia. Likewise, West Siberian producers may direct their lighter crude to Tuapse, Russia on the Black Sea.

Given these four factors, it becomes obvious that location, destination, distance, and direction are all spatial parameters that can be visualized, analyzed, and evaluated. Although crude quality is an aspatial component, geography still dictates where it is sold.

In the case study to be discussed, GIS has been used for:

- Geographic visualization
- Charting
- Arithmetic calculations

- Statistical analysis
- Distance measurement

Supply and demand

In 2001, the crude oil refining capacity of Russian refineries outran demand by about 42%. According to Petroleum Argus, for the 32 major plants (excluding 10 minor facilities), nameplate capacity for Russian refineries totaled 306 million tonnes/year (2.2 billion bbl). Yet in 2001, refiners consumed only 178 million tonnes of crude oil. As a result, domestic oil prices remained depressed by as much as 80% when compared to Dated Brent.

To initiate the analysis, the author defined the question, "What domestic refineries will provide the best price under the given conditions of market accessibility and price from production sourced in the Romashkinskoye field?" (The Romashkinskoye field is located about 650 mi east of Moscow.)

In such a situation, it becomes important to find and analyze information based on:

- Accessible pipelines that can reach neighboring refineries
- The locations, owners, technical capabilities, recent throughput, and nameplate capacity of these refineries
- The relative weighted refining end-product value of these refineries
- The cost to transport crude oil to these refineries along the pipeline system
- Access to refineries that are close to major urban centers or industrial/agricultural centers regularly in need of feedstock—consistent buyers in any pricing environment
- Finding refineries or blending facilities that can technically handle a specific crude type (i.e., sour crude, condensate, etc.)

Mapping these entities in a thematic structure provides an inherently proper way in which to answer the question.

The next step consists of thinking through the most efficient and structured way in which to solve the problem. This became a nine-step process:

1. Decide which geographic themes (layers) will portray all of the marketing environments that affect oil sales (i.e., pipelines, refineries, ports, etc.)
2. Identify the sources of public, private, and internal data
3. Collect the most accessible, affordable, and easily handled data
4. Manually enter, import, scan, compile, and digitize the data
5. Build individual themes
6. Combine the themes and build a GIS project
7. Choose the appropriate spatial tools to analyze the themes
8. Evaluate
9. Make a decision

Data sources

Without question, one of the most important themes to be built included data based on the technical refining capabilities and outputs as provided by Petroleum Argus. This data had to be spatially linked to the geographic location for each facility through manual spreadsheet inputs and then imported into the program. Much time was spent finding *approximate* geographic coordinates for each refinery, port, and city needed to spatially locate and link the data within the GIS.

Two excellent sources included a GIS file provided by ESRI that contained more than 65,000 city locations. [ESRI shapefiles actually consist of three interlinking files: a main file, index file, and a dBase table (see www.esri.com).[13]] The United States National Imagery and Mapping Agency GEOnet server at http://164.214.2.59/gns/html/index.html also provided another key source of worldwide city coordinates. This allowed searches on a variety of spellings.

Another equally important theme included the Russian pipeline network with diameter attributes provided by Wood MacKenzie. This GIS-ready theme was easily converted from MapInfo *mid/mif* files into ArcView shapefiles, showing excellent interoperable data conversion capabilities from vendor to vendor. Additional data related to pipeline capacity and other attributes were added by perusing publications. These included *Neft i Kapital* (*Oil & Capital*), *East Bloc Energy*, *Oil & Gas Eurasia*, the *Oil & Gas Journal*, and various Russian websites and annual reports. An important source included hard copy maps published by the Moscow-based mapping enterprise Incotec. This company provides fairly complete geographic coverage of the Russian oil and gas sector through maps and atlases produced on a variety of scales.

Data collection, quality control, and a format-to-fit GIS require a lot of work. Nevertheless, the data is there. It is "simply" an issue of turning the data into information, and then refining this information to make informed decisions. Some of the tools used to analyze the data will be described in the next chapter. Among these are spatial queries, distance measurements, buffers, and shortest path/least-cost analysis.

To answer the question of where to sell crude oil on the domestic Russian market at the best price, the themes described above were formatted in ArcView 3.3. They then were analyzed with help of ESRI's extension Network Analyst and visualized to evaluate various sales destinations. (Although the author used ESRI's Arcview 3.3 in this study, there are other software applications readily available to do such work, including MapInfo, Intergraph, AutoCad, and so forth.)

Pipelines

Russia's pipeline network contains 46,900 km, or 29,000 mi, of trunk pipelines and 395 pump stations. The network also contains 868 storage facilities with a total carrying capacity of up to 12.7 million m^3.[14] According to long-range development plans out to 2010, another 9000 km of pipe may be added by 2010.

Figure 2–4 (Plate 2–2) shows a map of the Transneft major trunkline grid with denoted diameters, directions, and refinery locations. Radiating from the Romashkinskoye field near Almetyevsk, Tatarstan, there are 3 major systems consisting of a number of parallel pipelines. From these trunklines, left-hand and right-hand turns lead to other refineries.

The first system from Romashkinskoye runs northwest then west to 4 refineries located near Nizhniy Novgorod, Yaroslavl, Moscow, and Ryazan. Past Yaroslavl, the pipeline heads northwest past Saint Petersburg, where it turns into the Baltic Export Pipeline, terminating at the port of Primorsk.

The second system carries crude southwest to refineries in Samara then east to the Syzran refinery. From here, the trunkline moves oil to export markets in Europe. Along this route, just before Lipetsk, a dedicated line swings southwest to the Kremenchug refinery in central Ukraine.

The third system also begins in Samara but carries oil to refineries near Saratov and Volgograd, then onward to the Novorossiysk export port on the Black Sea. Parallel to this line, just past Saratov, a dedicated line carries crude to Lisichansk, Ukraine.

Finally, a minor pipeline worthy of mention runs due north from Romashkinskoye to a small facility located near Nizhnekamsk, Tatarstan. This facility was being upgraded to a full-scale sour, heavy oil refinery and may become a significant downstream contributor by 2007.

Summing it up, producers can access 10 domestic Russian oil refineries from the Romashkinskoye field. The question is, which of these refineries will provide the best selling opportunities?

Refinery product value

The utility of GIS is demonstrated with its ability to chart, label, and color-code attributes to identify trends, relationships, and attributes. Figure 2–5 (Plate 2–3) shows segmented pie-chart plots for gasoline, gasoil, fuel oil, and jet fuel sold in 2001. Note the dominant nature of fuel oil and gas oil.

The 32 refineries used in the database provided by Petroleum Argus produced 128.4 million tonnes of refined products, of which 36% was fuel oil, 37% gasoil, 21% gasoline, and 6% jet fuel. The intent of this map is to provide a quick means to evaluate each refinery's product portfolio. Light oil with very little sulfur should naturally be diverted to those refineries with a high percentage of gasoline. Heavy, sour oil should be directed to refineries producing mostly fuel oil.

Depth of refining, labeled next to each pie chart, provides another indication of refinery quality as it indicates the yield of light products. Those refineries with the highest depth of refining should in theory provide the highest prices for feedstock, since the value of the product output is greater.

Carrying forward with this analysis, Figure 2–6 (Plate 2–4) shows 2001 refinery utilization denoted with a weighted price average in $/tonne for the same data set. In this example, July 24, 2002 Northwest European fuel oil sold for $133/tonne, gasoil (diesel) for $208/tonne, jet fuel for $228/tonne, and gasoline for $266/tonne. By calculating the percentage sales volume of each product type then multiplying by the above, it was possible to calculate the average end-product value. This value is posted in the center of the pie charts. Whereas Figure 2–5 (Plate 2–3) quantifies refining output, Figure 2–6 (Plate 2–4) quantifies product value. (Even the chart size conveys information concerning the relative sum volume of products sold in 2001.)

Again, it is those refineries with the highest end-product value that a producer or trader would target, since these refineries make more revenues per given unit of crude feedstock. In this example, the GIS calculated a minimum weighted price of $153/tonne, a maximum of $241/tonne, a mean of $196/tonne, and a standard deviation of $18.6/tonne for the 32 refineries.

To choose potential sales points, a simple GIS query quickly found those refineries that have an average mean output product value of $200/tonne or more. In the European part of Russia, within reach of the Romashkinskoye field, only 2 refineries met these requirements: Nizhnekamsk and Volgograd.

Yet it is important for producers to target not only those refineries with high end-product value, but those with some marginal excess capacity as well. Unfortunately, Nizhnekamsk's 2001 refining output capacity was low, and due to enforced price subsidies, this refinery was eliminated as an optimal choice. This left Volgograd with a weighted refinery output value of $200/tonne. Unfortunately, a Russian major company with substantial production in the area owns the refinery. As such, its relatively tight utilization in 2001 (87%), filled by its own production units, left the possibility of stable term contracts or spot trades questionable.

To obtain a secondary slate of sales options, it is convenient to utilize statistical analysis and move down one standard deviation from the mean. In this case refineries are sought out with a weighted refining wholesale price between $177/tonne and $196/tonne (see Table 2–1). This is accomplished by conducting spatial queries on attribute tables to select locations that fall within a certain range. Immediately, 8 refineries in Western Russia come to the top, including Moscow, Saratov, Samara, Nizhniy Novgorod, Novo Kuibishev, Ryazan, Syzran, and Yaroslavl. From these choices, Samara, Novo Kuibishev, Moscow, and Syzran provide the highest weighted refining sales prices at $190–$195/tonne. As such, these would be targeted as primary domestic sales opportunities.

Attributes of 2001 Weighted Price							
Company	Refinery	Location	Wt_price	Crude_k??	UH_01_%	Unum_cap01	Fuel_oil_k
Yukos	Achinsk	Achinsk	186	4979	0.77	0.23	2276
Central Fuel C	CFC	Moscow	192	9756	0.81	0.19	3094
Alians	Khabarovsk	Khabarovsk	184	2498	0.43	0.57	839
Sidanko	Creking	Saratov-Creking	184	3823	0.38	0.62	1382
Lukoil	Norsi	Nizhny Novgorod	184	6735	0.31	0.69	2523
Yukos	Novo-Kuibishev	Novo Kuibishev	195	6653	0.39	0.61	1599
TNK	Onaco-Orsk	Orsk	190	4054	0.51	0.49	1324
Lukoil	Permnefteorgsintez	Perm	195	10711	0.77	0.23	1824
TNK	Ryazan	Ryazan	188	10489	0.57	0.43	3966
Gazprom	Salavat	Salavat	186	5958	0.54	0.46	2142
Yukos	Samara-Kuibishev	Samara	195	5096	0.68	0.32	1565
Yukos	Syzran	Syzran	190	4711	0.45	0.55	1355
Bashneft	Ufaneftekhim	Ufaneftekhim	195	6651	0.53	0.47	1630
Slavneft	Yaroslavl	Yanos	187	11216		1.00	4190
Yukos	Angarsk	Angarsk	198	7240	0.31	0.69	1686
Gazprom	Astrakhan	Astrakhan	221	2145		1.00	373
Lukoil	Kogalymneftegaz	Kogalymn	227	167		1.00	0
Rosneft	Komsomolsk	Komsomolsk	166	4095	0.74	0.26	1562

Table 2–1 Russian Refinery Text-Based Spatial Query (source: Petroleum Argus 2001 Database)

If the cost of transportation, processing, end-product excise taxes, and commissions can be provided to calculate each refinery's netback, it then becomes possible to intelligently negotiate crude oil price with the owner. (Netback from a refiner's standpoint equals product wholesale price minus refining costs and feedstock price.)

Nelson Cost Index

To deal with the issue of finding the right crude for the right refinery, a similar spatial technique can be applied to visualize the Nelson Cost Index. (This is calculated yearly by Daniel Johnston for PennWell's *Oil & Gas Journal* Energy Database.) The cost index provides a qualitative indication of the amount of capital spent on a refinery. For example, refineries with catalytic crackers, hydrocrackers, and alkylation units are more expensive than units with simple distillation columns. As such, it is also an indicator of refinery sophistication. By plotting this data graphically across Europe and Russia, it becomes obvious that European refineries are more advanced than Russian refineries [see Fig. 2–7 (Plate 2–5)].

To add value to this analysis, Leo Aalund, an editor with the *Oil & Gas Journal*, uses the magazine's Annual Refining Survey to characterize refining yield. He simply sums the reforming, alkylation, cat cracking, and hydrocracking capacity and then presents it as a percent on crude distillation capacity. Through this analysis, Aalund has found that U.S. yields equate to about 75%; yields are 30% for Asia and 35% for Europe. In Russia, with its heavy output of fuel oil, refining yield on the average is even lower. In combination with the other reasons described, it becomes clearer why Russian domestic crude oil sells for about one-half that of world prices.

Another way to use the Nelson Index and other refinery characteristics supplied by the *Oil & Gas Journal*'s Annual Survey is to evaluate each refinery's technical specifications. Then specific criteria can be mapped to segregate heavy crude oil refineries from those that produce light products. Unfortunately, Russian survey respondents fall short of supplying the necessary information to make such an assessment at this time. But for those who wish to look at European and American refineries, the Nelson Index provides a wealth of information. For example, the Shell and Lyondale Citgo refineries located near Houston, Texas, use Mexican Maya and Venezuela feedstock. These refineries can be distinguished from other refineries by their high-capacity hydrotreaters.[15]

Far abroad

The final discussion in the chapter shifts away from domestic sales to far abroad transportation issues. Russian crude oil buyers typically pay for the cost of transportation, which is called free on board or f.o.b. to the seller. In contrast, the seller usually pays cost, insurance, and freight (c.i.f.) for far abroad sales. As such, far abroad transportation alone reduces income by 10–15% for each particular sale.

Figure 2–8 (Plate 2–6) shows two pipeline routes, both of which begin in Samara, but end in Novorossiysk and Slovakia, respectively. Along these routes, the GIS-calculated distance across a series of legs is depicted in km.

In turn, by knowing certain costs, it becomes easy to calculate the total transportation costs from almost any point within the network.

Information needed to ascertain transportation costs include Russian pipeline transportation tariffs, customs duty, commissions, Eastern European transit charges, and freight costs from published schedules. This becomes particularly important if a company is evaluating a new project and needs transportation costs to determine the rate of return, return on investment, and net present value.

During June 2002, the Transneft transportation tariff from Samara to Novorossiysk and Slovakia, respectively, was $4.78/tonne and $3.24/tonne.[16] It is 2164 km to Novorossiysk and 1419 km to Slovakia. With this information, and by using published schedules, the two can be compared. This will allow a determination of whether the cost to transport 1 tonne of crude over a distance of 100 km is the same for all of Western Russia. In this case, the correlation worked, and the cost to ship crude along both pipelines came to about $0.20 per tonne per 100 km.

Using this information, a company can evaluate, for example, a play that is situated at the beginning of Leg 2 along the Samara-Novorossiysk route [see Fig. 2–8 (Plate 2–6)]. It then becomes a simple matter of measuring the distance in GIS and multiplying by the transit cost. In this case, 1139 km times $0.002 equates to $2.28 per tonne. Subsequently, this information can be placed in an economic model.

Of all the tools in a GIS, distance measurement is one of the most fundamental utilities. It can be used for a variety of applications.

Conclusion

As can be seen, the study of geographic patterns and relationships forms the core utility of GIS. As a petroleum professional, if one wishes to determine the usefulness of GIS, one can ask:

- Is it beneficial to investigate statistical relations among phenomena in a given area?
- Is location delineation a common part of a company's decision-making process (well locations, pipeline corridor selection, or retail gasoline outlets)?
- Is there a need to analyze markets from a geographical point of view?
- Is the company heavily focused on subsurface analysis?
- Does the company analyze temporal and spatial changes in phenomena (the number of customers visiting a particular gasoline station during fall, winter, summer, and spring)?

If so, GIS may very well help in the decision-making process.

References

1. "GIS for Oil and Gas Conference Proceedings." 2000. *Ninth International Conference and Exposition, Geospatial Information and Technology Association*. Houston (September 18–20) pp. 227–233.
2. Heywood, I. et al. 1998. *An Introduction to Geographic Information Systems*. New York: Addison Wesley Longman. p. 96.
3. Jones, C. 1997. *Geographical Information Systems and Computer Cartography*. Essex, England: Longman. p. 89.
4. McDonnell, R. and K. Kemp. 1995. *International GIS Dictionary*. New York: Wiley & Sons. p. 33.
5. Heywood, p. 97.
6. Berry, J.K. 1995. *Spatial Reasoning for Effective GIS*. Fort Collins, CO: GIS World Books. p. 122.
7. Ibid.
8. Gaddy, D.E. 2000. "Journally Speaking." *Oil & Gas Journal* 98:34 (August 21) p. 15.
9. Ibid.
10. Ibid.
11. Ibid.
12. Shill, C. 2000. *Environmental Systems Research Institute's Petroleum Users Group Meeting*. Oral presentation. Houston (February 29).
13. ESRI Shapefile Technical Description: An ESRI White Paper. 1998. (July) p. 2.
14. Mikhailov, N. 2002. "Russian Oil Pipelines Set for Expansion." *Oil & Gas Journal* 100:12 (March 25) p. 62.
15. *Oil & Gas Journal Worldwide Report*. 2001. 99:52 (December 24).
16. "FSU Report." 2002. Petroleum Argus. 7:23 (June 14).

3

Spatial Fundamentals

We usually find oil in new places with new ideas. When we go to a new area we can find oil with an old idea. Sometimes we also find oil in an old place with a new idea, but we seldom find much oil in an old place with an old idea.

—Parke A. Dickey

The rest of this book now turns to the fundamental building blocks (models), spatial procedures (analysis), data collection tools, and other techniques needed to understand and use GIS. Because the text is oriented towards the layperson, it will not delve into the more onerous descriptions, in-depth calculations, or detailed flow charts needed to fully describe a GIS. Instead, the intent is to provide future users with an elementary knowledge of how spatial models and utilities work.

Fortunately, the casual user can begin to use GIS fairly quickly with positive results by doing a modest amount of homework. In addition to this book, suggested readings include Berry, Chou, and Heywood (see references). Additionally, a variety of on-line university programs can be found through UNIGIS (http://www.unigis.org), a worldwide network of GIS educational institutions.

Finally, many vendors like ESRI, Intergraph, and MapInfo provide on-line training or multimedia-based texts and tutorials tailored to their products. It is hoped that the reader will take advantage of these resources and begin applying GIS as a decision-support tool.

Let us begin now by describing the ways in which real-world features can be represented in computer models. This will be followed by discussions on basic GIS models, spatial statistics, topology, network analysis, multitheme operations, surfaces, databases, and interoperability initiatives.

GIS Models

GIS requires two integrated pieces to make a complete model: *spatial form* and *spatial relationships*. Whereas spatial form defines the structure and distribution of features in geographical space, spatial relationships focus on spatial interactions among these features.[1]

Keep in mind that a basic computer model does not know the difference between an oil well, pipeline, refinery, or a gas station. Instead, it looks at these entities as abstract bits and bytes, whose digital characteristics are fundamentally different than the real thing. But like any useful model, whether it is physical or abstract, a computer model does not have to replicate the real world in full. Instead, its function is to help recreate an *essential skeleton* so that the user will have a clear-cut, relatively cheap, and rapid means by which to solve a particular problem. By doing so, unnecessary information is deleted to deliver a more focused analysis. Simply put, the computer model strives to describe a complex world simply.

GIS analysis requires a different approach than what is generally applied to textual, casual flow, matrix, decision-tree, and other graphical (nongeographic) and numeric-based problem-solving techniques. Because of this, other approaches must be used to correctly format data and structure the solution. Nonetheless, the user should not discard traditional methods, but instead, these methods can be intertwined with new procedures to improve the analytical process.

One way to reorganize the spatial thought process is to look at the world through a child's eyes. When you take a stroll, do you focus on the relation of your footsteps with the ground? Do you examine the exact manner in which you circumvent curbs, potholes, and people? Probably not. As an adult, you perform this function without much thought. Children, on the other hand, look at spatial relations with due diligence. When they play jump rope, hopscotch, or jacks, they minutely scrutinize the interaction of body movements with the environment about them. Moreover, they constantly try out new ways of improving their play, such as moving up and down, or spinning clockwise instead of counterclockwise.

But as we turn into adults, we begin to lose our spatial sensitivities as we become comfortable with routine approaches used to solve everyday problems. In turn, this molds the way in which we perform our work. When asked to solve a particular problem, for example, we most likely turn to the tools and techniques acquired in college or on the job. If asked to determine reserves, for example, a geologist would probably use a volumetric approach, whereas an engineer would look at production curve profiles. Thus, even in our select professions, we get caught up in established norms.

Computer models have even become so much of a panacea for geologic mapping that fieldwork no longer plays the major part in investigative work that it once did. Even engineers forget to put down their calculators and spend time with operational personnel to see how original designs have been adapted to work more efficiently in the field.

The trick then is to try to recapture the way in which we once looked at the physical world and keep questioning how we look at objects and activities around us. To properly use GIS, therefore, one must closely look at the real world with a new perspective and trim it back to its essential structure and substance. As with any new technology, this is imperative, because novel applications always require a different approach for successful implementation. What better way to do this than with a child's inquisitive nature, guided by adult experience and education?

With this in mind, sit back and imagine the environment through which oil moves out of a reservoir to market. Fluids migrating out of the formation encounter permeability barriers, lithologic changes, preferential wetting surfaces, and void spaces. Hydrocarbons moving up the well encounter annular tubular restrictions, pressure differentials, and temperature changes. Fluids reaching the surface pass through phase separation vessels.

Gas and water being transported beyond this point may be injected back into the formation, while oil moves downstream to the local refinery. From here, distillation, alkalization, thermal cracking, and hydrotreating alter the chemical and physical structure of the crude to produce gasoline, diesel, fuel oil, and jet fuel. Finally, these products move towards various markets via pipeline, truck, ship, or railcar to filling stations, industrial facilities, or storage.

Now sit back and think about how you would describe the various natural and man-made environments through which these hydrocarbons pass. Each of us probably has some specific insights that would direct how we would look at this dynamic picture. Some methods would include mathematical computations tied to graphical representations (i.e., nodal analysis used in artificial lift design). Others would use 3-D representations based upon geologic analogs illustrated through structure and isopach contour maps.

Yet no single approach would be able to describe the movement of oil and the environment through which it travels. This is because as petroleum moves out of the reservoir into the distribution system, the spatial environment drastically changes, requiring a variety of disciplines to describe it all. So how can we make the computer program discern the differences among spatial phenomena, while attributing definitive information about each entity and the environment around it? As humans, we can use our senses to differentiate one from the other. A computer, however, cannot see, hear, feel, taste, smell, or sense time. As such, it must use a different method to analyze and portray abstract models of real world features.

Location and Attributes

In most GIS programs, a spatial entity (spatial form) is modeled according to its *geographic location* and the *attributes* used to describe it. First of all, geographic location defines the exact position of each entity, answering the question *where*. Accordingly, geographic operations focus on spatial distributions, geometric translations, linear and curvilinear measurement, coincidence, and spatial statistics. Secondly, attributes describe *what the entity is*, *what it does*, and *how it goes about it* (spatial processes). Correspondingly, attribute operations concentrate on theme building and layer construction. Thus entities can be queried, classified, aggregated, and combined into new layers through mathematical operations and conditional statements.

To unify the *where* with the *what* in a computer program, GIS models traditionally use three basic entities to describe the world: points, lines, and polygons. Cartographers developed these representations to portray 3-D features in a two-dimensional (2-D) format on a piece of paper. Spatial programmers, on the other hand, now use these entities as foundations for GIS models. (It should be noted that the object-based model, which focuses on describing assets and places, is slowly replacing the point-line-polygon model.) Unfortunately, determining which entity to use is not always straightforward.

Scale and shape

An astronaut looking at New York City from a distance of 150,000 ft might see only a point of light. But the same astronaut flying over New York City in an airplane would instead see a network of points, lines, and polygons representing automobiles, streets, and buildings. Thus observational distance, scale, and the relational interaction of one feature with another determine how a particular entity will be represented.

Consider the way oil fields are represented on maps of different scales. On a world map, for example, points best portray these features. But on a more local scale, say 1:10,000, a point would tell us nothing about the size and orientation of each field in relation to other fields. Obviously, a polygon would be used in this case. Yet at the same local scale, an oil well still would be represented best by a single point.

Move to within 200 ft of the wellbore, however, and you can imagine a line feature extending into the subsurface. As such, instead of a map view, a stratigraphic cross section portraying the interaction of the wellbore with the subsurface would be used. Move to within 5 ft of the wellhead, and the surface equipment clearly becomes polygonal in nature, in turn requiring a blueprint. Thus, depending upon your perspective and need, 3-D features in the real world can be represented as points, lines, or polygons in a computer model.

Scale, coupled with coordinate density (resolution), can also have a dramatic effect on form and shape. For example, at a scale of 1:15,000,000, the Volga River looks like a wiggly line [see Fig. 3–1 (Plate 3–1)]. [The city of Kamyshin in the upper view of Fig. 3–1 (Plate 3–1) is not even discernable.] However, at a scale of 1:1,000,000, one can begin to see obvious bends along its course. And at a scale of 1:250,000, meanders become quite pronounced. Finally, at 1:10,000, the choice of representing the river as a line or polygon presents itself through the inherent widening of the riverbank.

Attributes

As noted, attributes describe the basic substance of point, line, and polygonal entities. As such, they are used to define the unique characteristics of each object. Attributive descriptions can either be abstract in concept or concrete in nature. More importantly, they can be manipulated to garner new understanding about the object and the surrounding environment. One way to understand the significance of attributes is to examine how they add meaning to their spatial counterparts.

First, attributes can be *adjectives*. Just think of all the ways in which one can describe a well. Is it a gas well, oil well, or water well? Is it vertical, inclined, or horizontal? Second, attributes can be *nouns*. For example, is it simply a borehole or a well? Finally, attributes can also be *verbs*. Does the well "produce" or "inject" the oil, gas, or water? If so, how much does it produce in a day, a month, or a year? Thus, attributes tell how it is different, what it is, what it does, and how it compares with other features.

All this information, whether abstract or tangible, can be attached to spatial objects so as to lend new insights and understanding to our work. Again, it helps to scrutinize the world through a child's eyes and think about all spatial relationships, not just those that we have been programmed to identify. By doing so, a GIS will allow you to spatially relate an almost infinite number of attributes, taken from a variety of aspects, while using a variety of analytical methods.

From this point, how you use this information depends upon how you wrap your imagination around the multiple variables that can be mapped. As long as you can attach a descriptor to the spatial coordinates of the entity, a GIS model can be built. But what do you do with this data and information? Thanks to the science of statistics, it is possible to relate and analyze the properties of any entity with itself, with those like it, and with other dissimilar features around it.

Statistical Understanding

Statistics: the science of collecting, organizing, presenting, analyzing, and interpreting numerical data for the purpose of assisting in making a more effective decision.

—Mason and Lind

In 1951, S. S. Stevens categorized data into four levels of measurement (see Fig. 3–2):

- Nominal
- Ordinal
- Interval
- Ratio

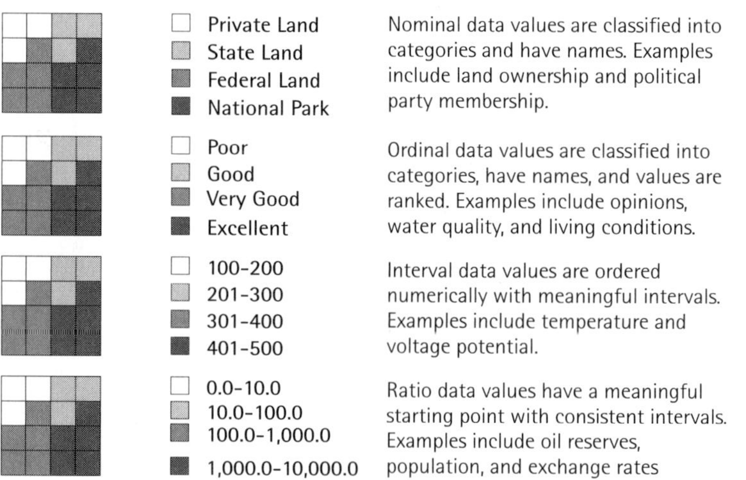

Fig. 3–2 Four Levels of Data Measurement

These categories play an extremely important role in working with GIS data as they allow the user to organize data types according to the intrinsic value of each measurement.

Nominal level

The term *nominal level* of measurement refers to data that are only classified into categories without reference to measurements and scales (see Fig. 3–2, top). Nominal scales are subdivided into two groups: renaming and categorical.[2]

Nominal renaming. Nominal renaming occurs when each object within a set is assigned a different number, that is, renamed with a number. Examples of nominal renaming include driver's license numbers or numbers on a lotto ticket. The former provides an ideal way to distinguish individuals, especially those with the same name (i.e., Randy Jones or John Smith). The latter allows the computer to easily handle thousands of entries utilizing an alphanumeric data set. The computer even can select a lotto winner randomly without regard to any predefined system of selection.

Nominal categorical. Nominal categorical measurement occurs when objects are placed into subgroups so that each object can be given the same number.[3] The subgroups must be mutually exclusive, that is, an object may not belong to more than one category or subgroup.

An example of the nominal categorical measurement occurs when grouping people into categories based upon political party preference (Republican or Democrat) or gender (male or female). In the former, Republicans might be assigned the number 1, Democrats, 2, and others, 3. In the latter, females might be assigned the number 1, and males, 2. In the nominal level, these arrangements can easily be changed. For example, males can be assigned the number 1 and females the number 2, or vice versa. As such, there is no particular order to the groupings.

Furthermore, the categories are considered to be mutually exclusive and exhaustive.[4] This means that a person cannot claim to be a Republican and Democrat at the same time. This also means that each individual, object, or measurement must appear in a category; in other words, it cannot be left out.

In the nominal level, standard statistical analyses including means, standard deviations, and correlation coefficients are most often impossible or impractical to use. For example, what purpose would it serve to define the standard mean of all Social Security numbers? One obvious exception occurs when the data has two, three, or more levels of nomination. For example, the designations could be: oil well = 1, gas well = 2, water injection well = 3, and water well = 4. In this case it is appropriate to both compute and interpret statistics as long as the unique requirements for each nomination are maintained.

It is not possible, however, to manipulate these numbers algebraically. For example, an oil well with a value of 1, and a gas well with a value of 2, does not make a water injection well with a value of 3; i.e., $1 + 2 \neq 3$. Clearly the nominal group does not provide the level of analytical value most GIS users require. However, in combination with other measures, it can be a useful tool.

Ordinal level

If we need to compare two or more objects, as we often do in GIS, we must move up one level. The ordinal level is a measurement system that possesses the property of *magnitude*, but not the property of intervals (see Fig. 3–2, second from top). One possibility might be ranking people in a church according to age. The youngest person is assigned the number 1, the next youngest person the number 2, and so forth. This is an example of an ordinal level.

Let's take a case where we are trying to find the most efficient way of traveling to a platform in the Gulf of Mexico. We could either:

1. Swim
2. Take a boat
3. Fly a helicopter

By ranking these three choices on an ordinal level from worse to best, we can rate them subjectively or objectively, depending on the ranker and the goal. Obviously, swimming would not be a good choice, and as such would be ranked on the worst end of the scale. And although a boat would most certainly reach our destination, a helicopter would be faster. On the other hand, a boat can transport more materials than a helicopter, thus providing a more practical way to carry pipe and equipment.

Thus, to make judgments on how to define an ordinal level, many factors such as opinions, form and function, economics, and time must be considered. As such, the major difference between the nominal and ordinal levels of measurement is the *greater than* relationship within the ordinal level. Otherwise, the ordinal level of measurement has the same characteristics in that it is mutually exclusive and exhaustive.[5]

As with a nominal level, GIS computations using the ordinal level may be less useful than precisely defined qualitative scales. If we need to improve the precision of our measurements, we must then turn to the interval level.

Interval level

The interval level is a measurement system that possesses the properties of magnitude and intervals, but not the property of rational zero.[6] It includes all the characteristics of the ordinal level yet maintains a constant size between each interval (see Fig. 3–2, second from bottom). The classic example of an interval level is the Fahrenheit temperature scale.

Suppose that three temperature measurements were taken in a wellbore: 100° F at 1000 m, 50° F at 500 m, and 0° F on the rig floor. These temperatures can easily be ranked, and we can also determine the differences between the temperatures. This is possible because 1° F represents a constant unit of measurement. But in this system, 0° F is arbitrarily chosen, i.e., it is just another point on the Fahrenheit scale.[7] Thus, 0° F does not represent the absence of heat; only that it is colder than 10° F.

Looking at the wellbore example again, although it may appear that the temperature recorded at 1000 m is twice as cold as 500 m, it is not. We can only say that it is 50° warmer. Just like the nominal and ordinal levels, the interval level form of measurement has mutually exclusive and exhaustive properties. The temperature on the rig floor, for example, cannot both be 0° F and 10° F at the same time. Thus, it meets the requirements of mutual exclusivity. Additionally, we can list the high temperatures for all days of the year. Therefore, it meets the requirements of exhaustivity.

Ratio level

Ratio levels possess all three properties of magnitude, intervals, and rational zero (see Fig. 3–2, bottom). One clear example is the metric ruler. It begins with 0, uses linearly constant graduated intervals for measurement, and can be mathematically sliced and diced. For example, if one cuts a 1-m long stick into two equal parts, each half will measure one-half meter, or 50 cm. Cut each of these pieces in half again, and then one has four pieces that are 25 cm in length. Now a comparison can be made with the temperature system. In this instance the concept of rational zero plays an important role in distinguishing between the interval level and the ratio level.

The added power of a rational zero allows ratios of numbers to be meaningfully interpreted. For example, the pie charts in Figure 2–5 (Plate 2–3) were calculated by dividing the output of gasoline, fuel oil, gas oil, and jet fuel by total production for 2001. Because total product volume in tonnes can be directly compared with gasoline volume measured in the same units, ratios and percentages can be used to compare output. As such, the ratio level is by far the most powerful measurement technique.

Now take a look at Figure 2–7 (Plate 2–5). The legend uses the ordinal level. If one took a look inside the GIS, however, one could see that ratio analysis actually determined the ranges of complexity based on graduated intervals of statistical standard deviations. Each level could have been categorized with some numeric identifier. Instead, the author strove to make it easier for the user to understand the graphical information by categorizing each subdivision from very low to advanced.

Groupings

The next step towards making a statistical spatial analysis is to classify data into appropriate groupings. All classification methods rank selected attributes from low to high values, then subdivide the ranked attributes into classes. In most GIS programs, the ability to classify data into groups is inherent. Thus, it is important to understand what tools are available so the user can look for patterns.

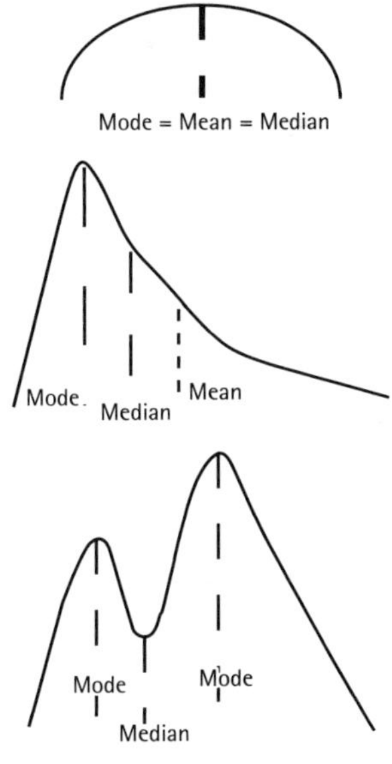

Fig. 3–3 Data Distribution Curves

Let's begin by looking at the top curve in Figure 3–3. It is symmetrical and bell shaped, meaning that the distribution has the same shape on either side of the center axis. In this type of symmetric distribution, the mode, median, and mean are all equally positioned at the center of the curve.[8] Thus, if the curve were to be folded in half, the two sides of the curve would be identical.

But as the distribution becomes asymmetrical (see Fig. 3–3, middle curve), or skewed, the relationships among the three statistical averages shifts. It then becomes necessary to evaluate the shape of the curve and divide it into sections so that data can be categorized in a logical manner.

Classification

In addition to using standard deviation in some GIS programs, there are other tools available to classify data. Some of these tools include *equal interval*, *quantile*, and *natural breaks*.

EQUAL INTERVAL. If a frequency data distribution consists of a relatively uniform pattern, then a simple equal interval breakdown (equal-step series) can be applied.[9] In this distribution, the difference between the high and low values stays the same for every class. Within the program, the GIS subtracts the lowest value of the data set from the highest, then divides this number by the number of specified classes. Next it adds that number to the lowest data value so as to obtain the maximum value for the first class. Finally it adds to each maximum value to set the break intervals for the rest of the classes.[10]

The equal interval frequency distribution provides an easy way to interpret data groupings because the range for each class is the same. This frequency distribution has provided great utility in mapping continuous data such as production rates and temperature. It also provides the GIS presenter with a useful means to present information in general terms. Unfortunately, sometimes the data values tend to be clustered rather than evenly distributed. If this is the case, too many features may be placed in one or two classes and too few in the others.[11]

QUANTILE. In the quantile classification, each class contains an equal number of features with objects arranged in ascending or descending order. According to Mitchell, the GIS orders the features from low to high based on the attribute value, summing the number of features as it goes. Next it divides the total by the number of classes specified to obtain the number of features in each class. The program then assigns the first features in the lowest class until that class has been filled, moving on to the next class and filling it in the same manner. This technique provides a good way to compare areas of roughly the same size. It also easily maps data in which the values are evenly distributed.

One problem with quantile classification is that data points with similar values can be forced into adjacent classes, at times exaggerating the differences between features. Conversely, widely ranging adjacent values may end up in the same class, minimizing the differences between these features.[12]

Natural breaks. Natural breaks classification organizes data into groupings of similar values. In this manner, data values that form clusters are placed into a single class. The GIS can break out classes by discerning gaps between clusters of values. As such, this technique provides an excellent way to group data values that are not evenly distributed. Unfortunately, because class ranges are specific to the individual data set, it is difficult to compare one map with another.[13]

Standard deviation. Standard deviation provides a method to measure the amount in which values in a population vary from the mean. If the standard deviation is relatively small to the arithmetic average, this tool can provide excellent insights for making decisions.[14] The GIS calculates the mean value by adding all the data values together, divided by the number of features. It then calculates the standard deviation by subtracting the mean from each value and squaring it. The program then sums these numbers and divides by the number of features.

The GIS can create class breaks above and below the mean based on the number of standard deviations specified by the user. This technique provides a way to measure and compare the spread in two or more sets of observations. As such, thematic similarities and differences from map to map can be compared.

Bimodal distribution. One final classification worthy of mention includes bimodal distribution. This distribution consists of two or more distinct groupings of values that have more than one bell-shaped curve (see Fig. 3–3, bottom). Often two points of concentration develop because the sampled population is not homogeneous.[15] For example, suppose the production rates for a sample grouping of wells are 20, 25, 35, 35, 35, 40, 42, 62, 67, 70, 70, 70, and 75. The two modes would then be 35 and 70.

Figure 3–4 shows how a textual representation of these numeric values does not provide the full picture. Take a look at the table values at the top right. After close and time-consuming scrutiny, two modes can be discerned. But by plotting these numbers, the graph at the top then easily shows there are two groupings. However, no true insights are provided until the values are spatially portrayed in a map view (see bottom map).

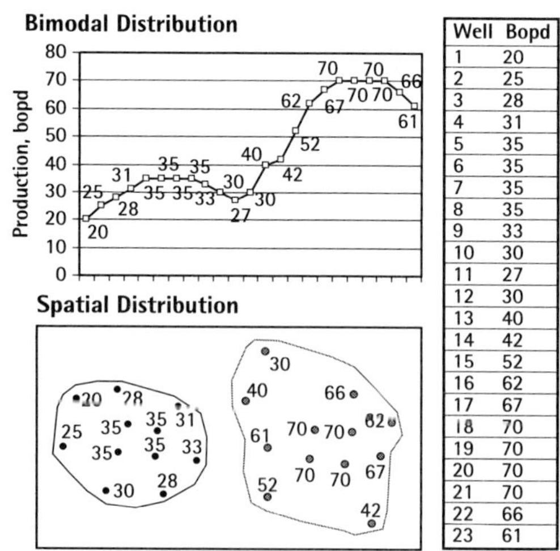

Well	Bopd
1	20
2	25
3	28
4	31
5	35
6	35
7	35
8	35
9	33
10	30
11	27
12	30
13	40
14	42
15	52
16	62
17	67
18	70
19	70
20	70
21	70
22	66
23	61

Fig. 3–4 Bimodal Chart and Map

In this example, the populations being sampled might provide an indication of the reservoir pressure and permeability for wells drilled on adjacent structures. Or they may indicate the maturity of field development. Unfortunately, from a purely statistical point of view, it becomes next to impossible to determine the meaning behind the statistics.

For a more useful interpretation, it becomes necessary to geographically place the data points on a map. Thus, one can begin to see how GIS can quickly improve interpretations by combining the old-world science of statistics with the newer science of spatial analysis.

Interpretation

Figure 3–5 (Plate 3–2) shows four density maps of senior citizens living in Denver, Colorado, using 2000 U.S. Census data. Although the base data are the same, partitioning values among four distributions clearly shows how interpretations can be affected by the end product.

For example, let's assume an oil company wishes to market gasoline credit cards to senior citizens. An equal interval classification would show only a handful of prime regional targets [see Fig. 3–5 (Plate 3–2), top left]. Thus of the four classes displayed on the map, the first class pretty much wipes out all spatial heterogeneity. (The first class calculates a density of 0–2010 senior citizens per square mile.) As such, there are no anomalies from which to build a marketing campaign.

On the other hand, quantile classification shows too many marketing opportunities. It unrealistically distributes senior citizens across three of the four classes [see Fig. 3–5 (Plate 3–2), bottom left].

Classification by natural breaks, however, depicts a high degree of spatial differentiation by distributing senior citizens more realistically located across the city [see Fig. 3–5 (Plate 3–2), top right]. Of particular interest is the class interval with a density of 823–2822, which shows several promising high-density areas. Standard deviation supports this view by showing a correlative overlay [see Fig. 3–5 (Plate 3–2), bottom right].

Of course, a more detailed investigation would include a point overlay that shows the location of all the company's gas stations across the city of Denver. Additionally, a density overlay of household income per area would help to find middle- and high-income areas. Residents in these areas would be more likely to apply for a credit card. Such additions can be easily incorporated into the analysis as long as the data are available.

One way to quickly determine the best classification system is to make a bar chart of data values distributed across the entire range. Then it is possible to examine how values fall into groups.[16] This can be accomplished using a spreadsheet and charting program. Set the horizontal axis to read attribute values and the vertical axis to represent the number of occurrences. If the data are not evenly distributed, and there are gaps in between groups of values, it is best to apply natural breaks. By looking at Figure 3–6, for example, natural groupings of data separated from one another by gaps can readily be seen.

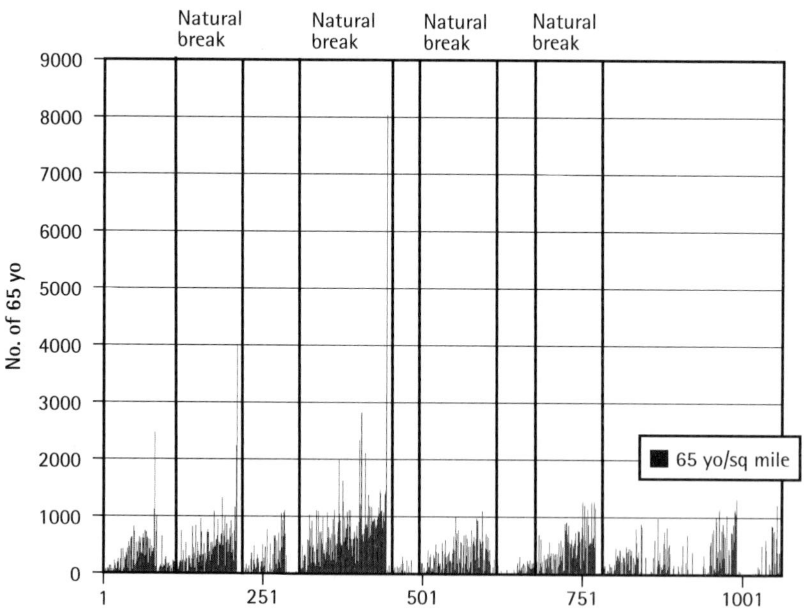

Fig. 3–6 Distribution Bar Chart

Now compare this numeric classification with the corresponding spatial classification shown in Figure 3–5 (Plate 3–2), top right. The chart shows a clear indication of the type of distribution to use, but the map conveys information that can be turned into knowledge.

In the GIS program, it is easy to change the classification, the number of classes, and the symbols. As such, spatial patterns from several approaches can be quickly visualized and interpreted even if one does not take the time to chart the data. When it comes to choosing the number of classes, four to five classes usually suffice. Too many classes will confuse the reader, while too few will not reveal much variation.[17] This shows the importance of experimentation so as to become fully familiar with the alternatives.

Statistical importance

The reason for this statistical discussion is threefold. First, it is important to understand how information can be categorized, especially in GIS. This way, one can decide how to structurally organize the data. Second, it provides the user with some understanding of the advantages and limitations that each categorical evaluation has on the end result. Third, it provides a powerful tool for interpretation.

Thus, the use of basic spatial statistics allows the user to intellectually select the best method for data treatment. If one decides to use an arbitrary classification system within the nominal level, for example, it would be impractical to use standard deviation, hypotheses testing, or chi-square.[18] Conversely, statistical methods provide an effective way to look at volumetric commodity data such as that provided for a variety of refineries. Further investigations will then make it possible to geographically compare the levels of light petroleum output of one refinery with another in relation to supply-and-demand issues.

The importance of statistical spatial analysis has often been overshadowed by other analytical functions. These include location-allocation modeling, site-selection, network-analysis, routing, and 3-D modeling for analysis. This should not be the case, as statistics can play a role in all applications.

Spatial Distribution

The main difference between classical statistics and spatial statistics is that classical statistics seek the central tendency (average) of data in numeric space whereas spatial statistics seek the map variation (standard deviation) in geographic space.

—Joseph K. Berry

We now turn to the topic of spatial distribution among point features. Point patterns have played an extremely important role in spatial analysis, especially in the petroleum business. For the past 50 years, geologists and engineers have refined development plans by simply looking at the spatial relationships between dry holes and producers. Furthermore, production trends that determine how many wells can be economically drilled, based on sound engineering practices, have led to state-regulated spacing requirements.

Moreover, the advent of more powerful desktop computers has facilitated the integration of GIS and geostatistics. This has changed the estimation and prediction methods used by earth scientists and engineers. Such integration can be used to assess naturally occurring phenomena whose values are in some way related by distance, area, volume, or space. As one expert writes, "in the coupling of GIS with geostatistics, what is envisioned is the use of geostatistical constructs to analyze point data contained in GIS coverages, or conversely, the use of GIS can improve geostatistical estimates from various information and/or to portray them as GIS coverages."[19]

Geoscientists have successfully used geostatistics to estimate rock and fluid properties located away from the wellbore. Furthermore, GIS complements the science of geostatistics by allowing the user to "simultaneously observe multiple data types of varying sizes." This permits "the discovery of relationships that might otherwise go unnoticed or that might require more expensive time and resources to detect."[20] There is no reason why this technique cannot be applied to other areas.

Point patterns

A point pattern is simply the distribution of a set of point features. In point pattern analysis, however, spatial properties of the entire body of points are studied as a whole, rather than as individual entities. Because points have no dimensions, valid measures of point distributions remain confined to the number of occurrences in the pattern and respective geographic locations.[21] Point feature distribution can be described by frequency, density, geometric center, spatial dispersion, and spatial arrangement. With the exception of spatial arrangement, evaluation of the spatial properties of point features again depends on basic descriptive statistics.

Geographic frequency. *Geographic frequency* refers to the number of point features that occur on a map. In most cases, the determination of frequency may be the first measurement to take place. It may also be the last, as changes in frequency over time provide great insights into the study of movements. If area is not considered, however, comparison of two distribution frequencies may mislead the interpreter. Thus, when two point patterns that differ in area are compared, it is prudent to evaluate patterns based on density.[22]

Density. *Density* shows you where features are concentrated. To calculate density, the program divides a value by the area of the feature to get a value per unit of area. By dividing the population of a county by land area in square miles, for example, you can obtain a value of people per square mile. Density does a good job of distinguishing distribution patterns when the size of the areas varies greatly.[23] For example, U.S. Census tracts have roughly the same number of people. Thus, smaller tracts indicate urban settings and crowded conditions, whereas larger tracts indicate a less densely populated and more rural setting [see Fig. 1–6 (Plate 1–1)]. However, when the comparative areas are nearly the same, other comparative tools must be used.

Geometric center. The *geometric center* of a point distribution, indicative of dispersion, is represented by an X-Y coordinate. Because maps have a 2-D format, the dispersion of a point distribution is related by the standard deviations of the X- and Y-coordinates along separate axes. In this definition, the X-coordinate indicates the dispersion of points along the X-axis and the Y-coordinate indicates dispersion along the Y-axis. Be aware that geometric centers do not provide a reliable indication of central tendency when either of the standard axial deviations have a large value. "Because the dispersion along the X-axis is independent from the dispersion along the Y-axis, it is possible for a distribution to show a large standard deviation on one axis and a small standard deviation on the other."[24]

Spatial arrangements. Point feature spatial arrangements form an important characteristic of a spatial pattern. This is because the locations of point features and relationships among them have a significant effect on the underlying process used to generate the distribution.[25] The three basic types of point patterns are:

1. Clusters. Point features that are concentrated on one or a few relatively small areas; forms easily recognized groups
2. Regular. Point features that are characterized by a uniform pattern of evenly spaced features that have a relatively large interpoint distance
3. Random. Point features that are scattered and have no apparent underlying design

Obviously, the human eye can readily discern the differences among patterns, shapes, colors, and textures. Unfortunately, a computer cannot do this without a little bit of help. Thus, a GIS model must be instructed exactly how to handle and display spatial objects. To make this type of analysis possible, let us move on to the topic of GIS data structures.

Vectors and Rasters

To make a model of the world, users can take advantage of the layer approach. For example, one layer may consist of an aerial photograph, the next topography, and the final land ownership. In most GIS programs, each layer is separately tiled. Such layers are clearly thematic and represent some unique feature (i.e., roads) or grouping of similar characteristics (i.e., transportation routes such as roads, railways, or rivers).

The primary GIS data structures used to conceptualize the world about us consist of *vectors* and *rasters*. The decision whether to use one over another will determine how the data will be organized, processed, and analyzed.

Vector data structure

The vector approach emphasizes the existence of discrete phenomena delineated by points, lines, areas, networks, and surfaces.[26] The simplest entity, a *point feature*, embraces all geographical and graphical entities positioned by a single X-Y coordinate pair.[27] (Point features are dimensionless, having no width, breadth, or height.) In addition to location, attribute data are then tied to the entity's coordinates to indicate what kind of point it is and what kind of associated data it contains.

Line. A *line* feature consists of a sequence of points connected to one another. Like point entities, attribute data are also tied to the line entity. Because a line vector has no width, it is a one-dimensional (1-D) entity. Obviously, many features have curvilinear form rather than a straight form. To replicate a curve, the vector data structure simply requires a high concentration of points to form arcs. The greater the density of points along the line segment, the closer the line approximates a complex curve.

Polygon. A *polygon* is represented by a series of lines that form a closed loop. Because a polygon has breadth and width, it is a 2-D form. In this case, the area of a polygon becomes a valid unit of measurement.

Surface. A *surface* contains all of the 2-D properties of a polygon, but includes a Z value that allows 3-D representations of features, such as elevation

Keeping these subjects in mind, vector topology, yet to be discussed, plays a very important role in how lines and polygons are treated in a vector-based GIS. These are used to produce networks and other interrelational structures.

Raster data structure

A raster data structure may be thought of as a computer photograph where each picture element (pixel) has a value that is associated with a geographic position [see Fig. 3–7 (Plate 3–3)].[28] In a raster structure, spatial features are organized in an irregularly spaced coordinate system. As a feature is decomposed into pixels, row and column positions reference each pixel to some portion of the Earth. In this case, cell length and width are fixed and the same. Thus, a pixel spans a cubic or rectangular area.

The cell values of rasters can be rendered in many ways. In a monochromatic image, shown in the top of Figure 3–7 (Plate 3–3), each cell has a value of 0 or 1, representing white or gray colors. In a color image shown at the bottom of Figure 3–7 (Plate 3–3), arbitrarily coded values are matched to precisely defined C-M-Y-K values.

Each pixel cell contains some representative attribute, such as height, precipitation, or land ownership. Because spatial features are geographically referenced, they can be drawn on the computer screen or used to perform an overlay analysis (see Fig. 2–2).

One of the main advantages of a raster structure is its ability to code each pixel with nominal, ordinal, interval, and ratio values. This allows map overlays to be mathematically related to one another. Referring to Berry's landslide prediction model in chapter 2, one can see that the raster data have great versatility. This is because various statistical levels can be incorporated into an analysis. For example, a weighted importance of the three layers—elevation, soil, and vegetation—can be cross-referenced across the cell attributes.

In one example, steepness might be identified as 5 times more important than either soil or vegetation cover in estimating landslide potential.[29] Referring to Figure 2–2, the weighted average would then be $[(9 \times 5) + 3 + 3]/7 = 7.29$. With all the layers built in the same coordinate system, it becomes relatively easy to reference one theme with another.

Mathematical operations, therefore, can be conducted on two or more raster themes to produce a final raster output layer. Moreover, functions such as +, -, /, *, Sqrt, Log, Exp, Ln, Sine and Cosine, and Boolean logic can also be performed on two or more rasters. This can be used to create output rasters with true and false values. Operands include And, Or, Not, >, >=, =, <>, <, and <=.

On the other hand, the raster data structure has certain disadvantages, especially in terms of data storage and entity portrayal. In many cases, for example, map features do not always need to be coded across the width and breadth of a dense grid. Take the case of a very flat plateau with an area of 5 km^2. If the pixel size used to model the plateau were built upon a 1-m x 1-m cell size, then there would be 5000 cells of redundancy within the database. Obviously, this would consume a lot of space on the hard drive as compared to the more compact vector structure. Programmers have developed several data structures to reduce this inherent redundancy. These structures include chain codes, run length codes, block trees, and quadtrees.[30]

Thus, the resolution of the pixel size defines the preciseness of the entity being portrayed. As such, it must be correctly chosen. For example, Figure 3–8 shows a line in both vector and raster formats. Remember, vector data use X-Y coordinates to exactly locate points, lines, and areas. Raster data, however, use cell grids. In a raster grid, point features are stored as individual column-row entries; lines are composed of connected cells. Areas are then composed of a patchwork of adjacent cells. Whereas the line vector is perfectly straight, the raster vector zigzags.

Figure 3–9 shows a similar portrayal of a contour map. In this case, pixel sizes determine the resolution of the output. Note the two linear contours in the northwest portion of the map. The vector lines are fairly precisely defined. The pixel sizes, however, are large; if used in an analysis, incorrect computations would result. Accurate computations and processing of spatial features therefore may not produce the desired effect if the proper pixel size is not chosen. To portray information obtained from a scanned map, satellite image, or photograph accurately, very small pixel sizes are often needed.

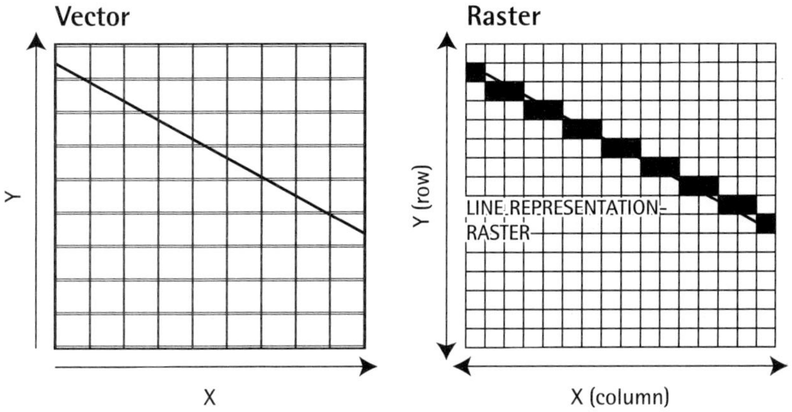

Fig. 3–8 Vector vs. Raster Data

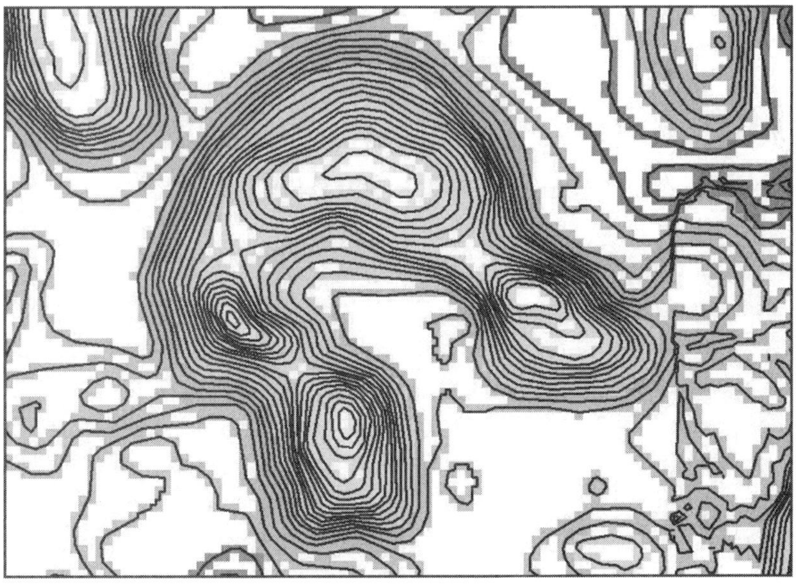

Fig. 3–9 Contour Vectors and Rasters

The need to increase resolution also introduces a need for more storage capacity and processing power. Fortunately, this is not much of a problem anymore with large hard drives and faster CPUs. Improved technology has provided an important turning point in the use of raster data. The petroleum industry already has collected an enormous amount of raster data in the form of satellite imagery, aerial photography, and scanned blueprints (see chapter 7). Geophysicists and engineers, moreover, have long used aerial photographs as backdrops for work involving the location of seismic surveys and pipeline corridors. Consequently, the use of raster data structures will grow.

As stated, map overlays are an easy task for raster data structures, but for vector structures, there are certain difficulties. For instance, determining whether a point feature is located inside a raster polygon can be a straightforward procedure. This is because all you need to know is whether the coordinates, as indicated by row and column, are part of the coordinates pertaining to the polygon.[31] The same procedure in the vector structure, known as a point in a polygon, a line in a polygon, or a polygon within a polygon, requires more complicated computations.

Vector Topology

Earlier in this chapter, we briefly discussed how differently humans and computers visualize the world. Humans see in color, readily distinguish differences in line width, and even sense textural variations. A simple computer program used to make maps, on the other hand, only sees digital numbers. In other words, a cartographic program only senses thousands of on-and-off dots and thus has no idea what it is rendering.[32] This leads to a very important note. Different computer programs look at data differently. For example, a rudimentary computer-mapping program remains blind to spatial patterns.

A GIS vector program, on the other hand, utilizes topology to determine and measure *proximity, connectivity, containment,* and *adjacency.* It allows the program to determine what kind of entity is next to, within, and away from other entities. And it allows users to determine relationships. The study of such relationships is a very important part of GIS as it allows users to track and manage the movement of energy, information, and commodities.

The building block in vector topology consists of the line entity. Simple lines by themselves carry no inherent spatial information about connectivity. By applying *topological pointers* to the data structure, however, it becomes possible to achieve a linear network that can be traced by the computer from line to line.[33] The programming language used to describe a topological structure varies from program to program. The following description is a compilation of commonly applied terms.

The pointer structure built in a GIS network model utilizes *nodes* that define *edges (links)* and *junctions.* Some examples of edges are streets, transmission lines, pipe, and stream reaches.[34] Some examples of junctions are trunkline tributaries, street intersections, and switches. Edges connect at junctions, transferring flow from one edge to another.

Nodes have two sets of numbers: a pair of coordinates and an associated node number. The node is much more than an adjoining point. It is an intersection of two or more lines that signal the start and end points of edges.[35] Again, edges are nonintersecting lines between points, representing the position of linear features or a bounding portion of a polygon. Consequently, identification codes can be used to answer queries or perform analyses. Some GIS programs, for example, track left then right along the edges to see which polygons are adjacent along its length. "This design feature allows a computer to know the actual relationships among all its graphics parts."[36]

As such, topology is not concerned with the physical shape of the entity, but the connectivity between entities.[37] One way to visualize this is to compare a subway map of New York City with the same underground tunnels placed on a city street map. Typically colored lines of the subway map show unidirectional segmented line paths. Intermittent stops used by passengers to gain access into the city (stop node) or transfer stations to another train (transfer nodes) are also depicted. Inherently, both distance and direction between nodes cannot be measured on a subway map. On a city street map, however, direction and distance can be portrayed with some accuracy, although topological interactions are difficult to formulate.

Fortunately in a GIS, the properties of both connectivity and geographic relationships can be maintained. Various GIS programs utilize different topological structures. Nevertheless, all topological structures try to:

- Remove node and line segment duplication
- Allow cross-references of line segments and nodes with more than one polygon
- Ensure all polygons have unique identifiers
- Represent island (i.e., a forest within an open plain) and hole polygons adequately[38]

There are several topological models in use, such as the geographic-based file/dual independent map encoding (GBF/DIME) model. The U.S. Bureau of the Census originally created this model to automate the storage of street map data in the 1960s.[39] This work led to the development of the well-known TIGER system used in the U.S. Census. The ability to identify relationships among nodes and edges allows the user to apply network analysis in problem solving.

Networks

A network is a set of interconnected linear features making up a set of features through which resources can flow.[40] As such, network models in GIS are abstract representations of the components and characteristics of their real-world counterparts. In the network model, several parameters determine how the program models activities so that the user has a useful tool for decision making. These include *nodes, links or edges, stops, centers, connectivity, directionality*, and *impedance*. The data regarding these characteristics are stored as attribute information in the vector model database.

In some definitions, *links* consist of arcs (complex lines) that represent pipelines, roads, rivers, railways, or power lines. *Nodes* are the endpoints of network links or edges. As previously stated, they represent switches and valves in utility networks, junctions in transportation networks, and confluences in streams. *Stops*, therefore, are node locations along a network that can be used for a variety of activities such as intermediary storage for alternative sales points. *Centers* are discrete nodal locations from which resources are derived, such as an oilfield or refinery.

Turns represent the transition from network link to network link at a network node.[41] Turns often dictate the terms of connectivity and greatly affect movement through the network system. For example, turns at a major tank farm or crude processing facility will slow down the flow of oil through the network. This occurs not only because it is a stop, but also because oil is often diverted to different destinations from this point. On the other hand, some turns are prohibited as flow directionality may provide only one path. Connectivity therefore provides information on the way networks join at network nodes and defines the direction in which materials move along the network.

An important element of network analysis, especially in economic evaluations, includes the ability to define *impedance*. Impedance, often referred to as frictional resistance along a particular flow direction, can represent costs and risk associated with traversing a network link. Impedance values, for example, can correspond to tariffs, low oil prices, weather events, high transportation costs, demurrage risk, and so forth.

Each link in turn will have different impedance values depending on local conditions. For example, a storm system in the Black Sea may force the seller to divert a shipment to an alternative destination. Impedance values are therefore important in determining the outcome of route selection, allocation, and spatially interactive operations.[42]

A network can be quite expansive and can include infrastructure beginning at the wellhead and ending at the gasoline pump. As such, several alternative paths may have to be analyzed before constructing the route with the least cumulative impedance.[43] Figure 3–10 shows how part of the map in Figure 2–4 (Plate 2–2) would be represented as a vector network diagram (refer to chapter 2, the Russian marketing example).

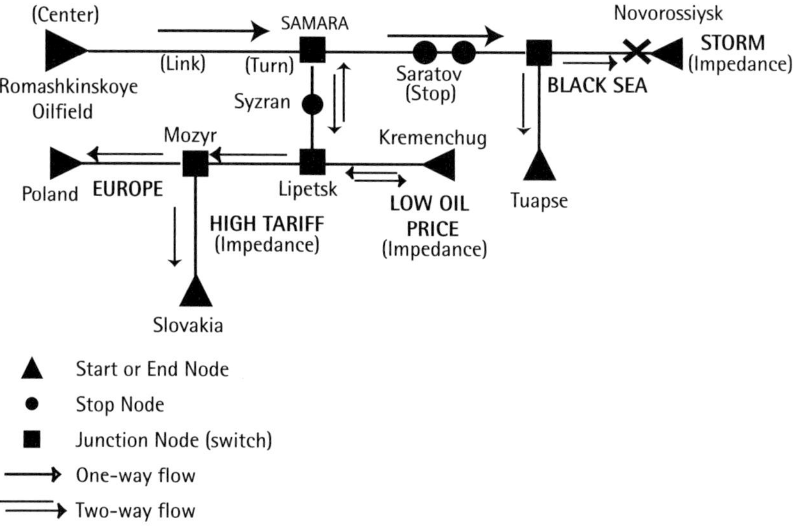

Fig. 3–10 Vector Network Model

In this case, the starting point would be the Romashkinskoye oilfield (center). From here, crude oil is pumped to Transneft preparation facilities at Samara in order to bring oil quality up to Urals specifications. (This results in time delays.) Then the oil passes through a junction node (switch) that directs the product to the next destination. Along the way, there are stop points in the domestic market where oil can be alternatively sold before reaching the Black Sea or Europe (i.e., Syzran).

Thus by looking at Figure 3–10, arcs become network links that represent pipelines, roads, air routes, and so forth. Along the way, these links intersect nodes. These nodes can either be stopping points (domestic refineries), intersections (Transneft storage and trunkline switches) or end points (European refineries or export ports). In the model, impedance values can be included along the links and nodes to guide the GIS user in selecting the best sales destination. Low oil prices at Kremenchug, for example, may suggest that the oil should be sold to another destination. Obviously, to build a more accurate network model, impedance values may be based on fiscal systems.

Shortest path

A typical problem in network analysis includes finding the shortest path from one node to another through a network. Because data in a GIS are organized in a digital format, such questions can be answered through the help of algorithms.[44] An algorithm contains a set of mathematical expressions or logical rules that can be followed repeatedly to find a logical solution to a question. To determine the shortest path, algorithms such as *minimal spanning trees* can be adopted to find the optimal solution.

A minimal spanning tree connects all nodes in the network with a minimal number of edges. This criterion first requires that the number of edges is equal to one less than the number of nodes, with $e = n - 1$. In this case, e denotes the number of edges and n denotes the number of nodes.[45] Secondly, the route for every tree must be located at one of the nodes in a network. Therefore, the number of trees that can be constructed must be the same as the number of nodes in the network. Construction of a minimal spanning tree requires the same number of iterations as there are nodes.

To find the shortest path from the origin node to the destination node, the analyst follows the procedure of tree building rooted at the origin node. The procedure starts with finding the first and closest node to the origin. It then continues on to find the next nearest node. In each iteration, the tree grows another branch. As soon as a node is reached (a result of this additional branch), a logical condition is evaluated to determine if the current node is the destination node. If the condition fails, the procedure moves to the next iteration. When the destination is finally reached, the algorithm terminates because the shortest path has been found. The shortest path algorithm can solve a wide variety of transportation problems, such as delivering gasoline to filling stations from a refinery.

Traveling salesperson

Another typical problem easily handled by GIS is the *traveling salesperson* problem. This name arises from the need for sales personnel to visit a series of clients within a given day. To save time and money, this must be accomplished by choosing the best path. In some cases, this may not necessarily be the shortest path. Problems with construction, traffic congestion, and poorly timed traffic lights, for example, may make alternative routes a better option. Consider a case in which a salesperson must visit five customers. The possible number of travel combinations in this case becomes five factorial (5!), or 120 alternatives.

In GIS network analysis, the ordering of the stops can be determined by calculating the minimum path between each stop. This can be considered for every stop on the list based on impedances set in the network.[46] Then a trial-and-error method can be used to order the visits so that the total impedance from the first stop to the last is minimized. The impedance level can be modified to take into account obstructions that can slow down the salesperson's driving time.

Resource allocation

Location-allocation modeling network analysis also can be used to allocate resources by modeling supply levels and demand levels throughout a network. To match supply with demand requires the movement of goods, people, information, and services through the network. In other words, levels of supply must be balanced with levels of demand. Allocation methods usually work by allocating links in the network to the nearest supply center, taking impedance values into account.[47]

Values for supply and demand can also be used to determine the maximum catchment area of a particular supply center, based on the demand located along adjacent links in the network. Without regard to supply or demand limitations, a given set of supply centers would service a whole network. If limits to supply levels or demand levels are indicated, coincidence situations can arise where parts of the network are not serviced, despite the demand being present.

All of these network procedures, and others, require fairly complicated algorithms. If the reader requires more technical information, research of many of the publications provided in the references may prove helpful. (Some useful resources are provided by DeMers, Zeiler, Burrough, Chou, Korte, Heywood, and Berry.)

Topological Operations

Although we have covered the topic of multiple layer operations to some degree, there are a number of tools in GIS that allow the user to generate new information from preexisting data. Multiple layer operations, also known as vertical operations, are based on the logical relationships among data layers.[48] These operations provide essential tools for spatial analysis because the user can manipulate data that have been organized as distinctly separate themes.

The user also can examine relationships among different features. Thus, it is possible to separate data on the same layer into multiple layers and then individually analyze related elements of spatial phenomena. Moreover, multiple layers of geographical features that cover the same area can be combined to form a single layer. This allows effective processing and model building.[49] Such multiple layer operations can be classified as *overlay*, *proximity*, and *spatial correlation* analyses.

Overlay analyses

Overlay analyses manipulate spatial data organized in different layers to create combined spatial features according to logical conditions specified in Boolean algebra.[50, 51, 52, 53] The logical conditions are specified with operands (data elements) and operators (space relationships among data elements). Operators include *and*, *or*, *xor* (exclusive or), and *not*. Each operation is characterized by specific logical texts of decision criteria to determine if the condition is true or false.

A core competence of a GIS program is the ability to edit and create new themes from two or more input themes. Topological operations allow users to *combine* (unions), *merge*, *clip*, and *intersect* two or more themes to create new themes. Such editing functions can be used to perform point-in-polygon, line-in-polygon, and polygon-on-polygon operations.[54] All four operations merge spatial features on separate data layers to create new features from the original coverage.

Variations on attribute retention and geographic extent depend on what Boolean operations are applied. In some editing functions, for example, the geographic extent of both input themes may be kept so as to only keep a portion of the attribute information (merge). On the other hand, attribute information of both input themes may be retained, but only the geographic extent of the island theme is kept. Other iterations along these lines are possible.

UNION. The union procedure, for example, uses the Boolean *or* operation to combine every feature on every layer into one output coverage. In other words, it appends the features of two or more features into a single layer and retains all attributes to the full geographic of extent of both themes.

Merge. In contrast, a merge procedure does not keep all of the attribute information from both input layers. Instead it consolidates features by merging them based on their values for a particular feature.[55] As such, only those attributes from the parent theme will be retained in the output unless the child theme contains the same fields.

Let us take a case in which attributes from Table 1/Theme 1 contain data fields for well numbers, production rates, and pump types. Table 2/Theme 2 attributes, on the other hand, contain fields for only well numbers. The merger between the two would eliminate production rates and pump types, while keeping the wells.

Another example is shown in Figure 3–11 (Plate 3–4), where federal lands in Washington are merged (combined) with urban lands. As can be seen, all of the parks, bureaus, and other lands have been reduced from 10 themes [see Fig. 3–11 (Plate 3–4), left] to 1 theme [see Fig. 3–11 (Plate 3–4), right]. Thus the merge feature in GIS programs combines features for 2 or more themes into a single layer. In this case, every polygon in the output coverage carries only attribute information from the urban lands theme.[56]

Clip. In the clip operation, the outline of the clip theme controls geographic output. The program only saves attribute information from the input theme and the geographics extent from the clip theme. This operation works similarly to a cookie cutter, with the final output based on the shape of the cookie cutter.

Intersect. The intersect procedure uses the Boolean AND operation to produce an output theme with features that have attribute data from both themes. Similar to the clip operation, the areal extent of the clip theme controls the geographic outline. However, attribute data from both themes are retained within the confines of the clip theme. Thus, when the two coverages are processed, only the portion of the input coverage that falls inside the intersect coverage will remain in the output coverage.[57] While the intersect coverage must be a polygon coverage, the input coverage can be a point, line, or polygon coverage.

BUFFER. One distinctly different, yet very important, topological operation is the buffer. A buffer is a polygon created through reclassification at a specified distance from a point, line, or polygon entity.[58] Buffering is simply a matter of measuring a distance from another object in all directions from that point. For point features, the GIS typically measures a uniform distance in all directions from that point, whether it is a raster or vector model.

For line and polygon features, however, in many cases the buffer distance will not be the same along the entire length of the object. Variable buffers, for example, can be created by using different impedance values on either side of a line or face of a polygon. This type of operation can be quite useful for specifying a radius of drainage around a wellbore. It is also useful for identifying the number of oil wells along a pipeline corridor.

Surfaces

As is often the case in the upstream petroleum industry, entities may also require the third dimension, or a Z parameter, to define exact positions on the Earth. In order to model the top of a hydrocarbon-bearing formation, for example, the elevation of the rig must be recorded. Only with this data can well logs be used to calculate the true vertical depth of a particular horizon. And only after this calculation does it become possible to calibrate spatial depth markers with seismic time markers to generate a contour model of the subsurface.

Ironically, in 2-D map form, this contour model is portrayed by a series of lines (see Fig. 2–1). Thus, what is actually a subsurface polygonal feature, which most likely has been folded, fractured, or faulted, is illustrated as a series of lines.

There are numerous ways to model a surface with both vector and raster techniques including tesselations, digitized contours, Z-value interpolation, and digital elevation models (DEM). Rasters representing surfaces as elevation data are widely available from sources like the USGS. Raster modeling can also easily perform spatial analysis such as dispersion, coincidence, proximity, and least-cost path. Such models have some disadvantages, however, because surface continuities like cliffs and folds are not well represented. Other features, such as peaks on mountains or gas caps in a reservoir can also be lost in the grid sampling. As such, vector topologies are preferred when topographical, structure contour, or other surfaces that utilize variable slope data are required.

The following is a description of one such vector model, the triangulated irregular network (TIN).

TIN

Triangulated irregular network refers to the formation of an optimized set of triangles based upon some defined set of points. Triangles make excellent representations of localized surfaces because three X-Y-Z points with values of spatial locality can uniquely define a plane in 3-D space. One advantage of a TIN is the ability to sample irregular distributions of points with variable density to efficiently model areas where the change in slope is abrupt. The term network in the abbreviation refers to the topological structure that is implicit in a TIN.

TINs represent surfaces as contiguous nonoverlapping triangular faces.[59] Surface values for any location can be interpolated through binomial computations of elevation within a triangle. Due to the unique properties of triangular geometries, triangles cannot be divided into subdivisions without creating additional nodes. The third dimension, or Z variable, can be placed at the vertices of the triangular facets such that both slope and aspect (slope orientation) of the facet are computable.[60]

Because elevations are irregularly sampled, the user can apply a variable point density to areas with steep slopes, producing a more accurate surface model. Thus, areas with wide variations in terrain require higher data density values so as to more accurately represent relief. The reverse holds true for terrain that has uniform slopes or is flat.

GIS programs often apply the Delaunay triangulation algorithm to optimize surface modeling with the structure model regarding nodes of the network as primary units.[61] Topological relations are then built into the database by constructing pointers from each node to each of its neighboring nodes. The basic idea of this algorithm is to create triangles that are collectively as close to equilateral shapes as possible.[62] Once the computer runs the algorithm, the TIN stores a list of nodes for each face, and for each face, a list of neighboring faces. This representation is similar to linear topologies discussed previously, except that the nodes have elevation attributes or attributes like formation thickness. The faces form triangles instead of arbitrary polygons.

Contours and grids

Other widely used methods in surface contour modeling include interpolation, gridding, triangulation, and kriging, mostly applied in the upstream part of the business. Some of the techniques provide straightforward estimation of distance based on the ratio of the distance between the two closest points of two adjacent contours. Unfortunately, this only works well with constant, regular, and continuous slopes, because values organized as irregularly spaced points do not always provide optimal distributions for spatial analysis. The process of gridding, however, does provide a means of turning these irregular points into a spatial pattern of high order.[63]

Two grid generation techniques include *local estimation* and *global approximation*.

LOCAL ESTIMATION. Local estimation means that the Z-value of a specific point location is estimated from the data of its neighborhood. In this method, the user must decide on the technique used to select the number of points for estimation. Local search methods include nearest neighbor, quadrant, octant, and various radii. The nearest neighbor method in particular, which confines the search to a given measure of distance, is reportedly the most commonly applied gridding mechanism in GIS.[64] For most applications, this method provides adequate results.

When the surface is systematically patterned, however, the search pattern may not ideally choose the optimal grouping of data points. This occurs because some values that obviously should be included may be excluded. This problem is analogous to difficulties with equal interval distributions described earlier in this chapter. Similar problems arose in which data points that would obviously improve the interpretation were left out of the analysis.

GLOBAL APPROXIMATION. In comparison to local estimation, global approximation uses the entire array of available data. The most commonly used approaches are polynomial trend surface method and kriging. For geostatistics, kriging reportedly provides the optimal method for spatial linear interpolation.[65] This technique uses two factors: the drift in regional tendency and the residual of local fluctuation. Some authors feel that kriging offers the most accurate method for spatial interpolation. For specific information on polynomials, regressions, variograms, and other formulae used in surface interpolation, please refer to Burrough and other authors listed at the end of the chapter.

Building a Model

The next discussion in this chapter concerns the four stages of model building that provide a structured guideline for organizing a project. When a GIS user needs to build a spatial model to analyze the distribution of phenomena, there are several things to consider. The GIS user must consider the nature of the model, the definition of geographic units, the specification of explanatory variables, and the quantitative method to employ in the work.[66] Spatial models can be divided into four types:

1. Descriptive
2. Explanatory
3. Predictive
4. Normative

In theory, each spatial model occurs in consecutive order, but in practice, they are typically interdependent. "After all, it is impossible to explain the distribution of any phenomena unless one can describe the distribution, and is impossible to predict spatial patterns unless one can explain the effects of significant factors."[67]

Descriptive models. In the descriptive stage, a GIS analyst strives to characterize the distribution of spatial phenomena. As such, description deals with spatial order and is represented by a set of descriptive statistics and indices. This might include the density of refineries located within a particular region that serves a local population. In general, descriptive models provide the basis and needs for other stages of model building.

Explanatory models. In the explanatory stage of modeling, the user deals with spatial association or relationships between the phenomenon and the factors affecting the distribution. To explain why some pipelines lead to export markets, for example, while others serve domestic markets, require many considerations. These include the role of supply and demand, politics, topography, and economics.

Predictive models. Once the distribution of spatial phenomena has been explained, *predictive* models can be constructed.[68] Predictive models can both forecast and simulate alternative management strategies. In the former, current conditions and extrapolations can be used to predict what will happen to phenomena at some time in the future. In the latter, users can develop scenarios based on varying certain conditions.

Normative models. The final (yet most complex) type of model building takes place in the normative stage. This activity requires different types of modeling such as linear and nonlinear programming. The general procedure for building a normative model includes the specification of an object and the identification of constraints.[69] In a typical transportation planning problem, the objective function may be to minimize total transport cost, subject to balanced constraints. These constraints take into account the need to equalize the total volume shipped out from the production facilities and the total volume received at each destination.

Relational data model

Obviously, no conversation on data models would be complete without a discussion on database models, of which the relational data model is the most widely used. The relational model represents all data in the database as simple 2-D tables called relations. Each table contains records for one entity or object. The tables appear similar to flat files except information from all related tables can be easily queried, extracted, and combined. Table structure consists of rows, which are called records or tuples; and columns, which are called fields. Each field has a distinctive topic and contains the attributes or coordinates for each entity.

Data entered into any cell down a field must obey rules established within the database structure. These established domains can be integer, value, date, or text-string based.[70] Thus, a text string cannot be entered into a date string within a field. Furthermore, there can only be one entry per cell. Null values can be applied for unknown data values, and each record must be distinctive from others.

In many cases, data from a number of relations can be used to produce a report. Herein lies the strength of the relational model: "It can relate data in any one file or table to data in another file *as long as both tables share a common data element, or key*."[71]

Figure 3–12 shows how it is possible to link spatial and attribute data in GIS using a simplified relational data model. Again, as long as both tables share a common element, like a unique identification number (ID), then countless relations can be established. Additional tables can be added as long as there are interconnecting IDs or keys that can link each table to the others.

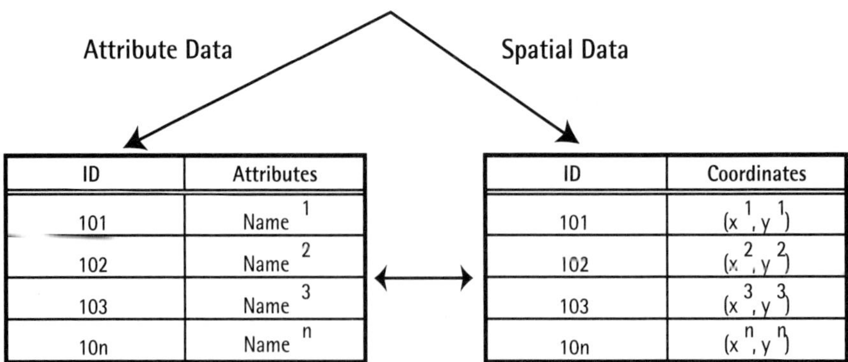

Fig. 3–12 Simplified Relational Data Model

In a relational database, three basic algebraic operations can be used to develop new sets of data:

1. Select—To select a subset of records
2. Project—To select a subset of fields
3. Join—To join tables based on key fields[72]

The relational data model has become a standard and integral part of many GIS programs. This makes it easy to add a new spatial entity (i.e., a new oil well) or attribute field (i.e., a new pay zone). It also provides an easy means to add such information directly into the table or even graphically by working within the geographic data extent.

SQL (standard query language) has been developed to facilitate relational database queries. SQL, however, has certain shortcomings with GIS, as it was not developed to handle concepts such as *near to*, *far from*, or *connected to*.[73] GIS software vendors have been overcoming this hurdle through further refinements in GIS database integration.

Object model

One move in this direction is based on logical object models rather than the standard entity relationships. One can think of an object as a person, place, or thing, rather than a point, line, or polygon. Thus, an object can represent a customer, refinery, or oil well. These objects are stored as records defined by attributes depicted along the fields. Furthermore, relationships among the objects can be established by using English language equivalents. These include: *is a part of*, *is owned by*, and *is located at*.[74] This type of logical database strives to bring the GIS user's model development closer to how it is perceived in the real world. Object-based modeling appears to be the next evolutionary step in GIS development.

Interoperability

One of the issues already mentioned in chapter 1 concerns the problem of interoperability. Interoperability is the ability for two or more systems to provide and accept services from each other and exchange data and information efficiently.[75]

The Open GIS Consortium (OGC) is an international industry consortium consisting of 250 companies, government agencies, and universities. It is undoubtedly the most involved organization in working towards interoperability in the GIS industry. Within this group, a large effort has been put forth by the petroleum sector. Shell and the European Petroleum Survey Group, for example, have made important contributions towards a comprehensive petroleum reference model.

OGC's mission has promoted interoperability in the information technology industry. Specifically targeted are geographic information systems, GPS, Earth imaging, facilities management, surveying and mapping, digital cartography, navigation, sensor/camera location, and location-based services.

OGC's Technical Committee is open to participation by all OGC members. Working groups review, develop, and vote on specifications for interoperability interfaces and encodings. Vendors and integrators now use OpenGIS specifications to make dissimilar software systems communicate directly in real time. Users usually do not need to convert one spatial data format to another.

In this way, two dissimilar systems using proprietary or special purpose internal formats can exchange instructions in real time through the *lingua franca* of their shared interfaces. As such, the network of linked computers acts as one computer, which then can be applied to geoprocessing applications. Thus, spatial data formats are no longer confined to expensive, full-function GIS programs. Instead, they now can be used and presented in many other kinds of applications.

Through the OpenGIS Web Map Service Specification (raster views) and the OpenGIS Web Feature Service Specification (vector data), spatial data can also be served to Web browsers. This allows automatic overlay of data to other Web servers. This includes the ability to automatically transform data and views to the same scale and coordinate reference system. For Internet-based applications, OGC members have created the Geography Markup Language. This is an XML hybrid that may become the industry's standard way of encoding spatial data.

Some interoperability initiatives in the early 2000s include:

- **Critical Infrastructure Protection Initiatives.** CIPI-1 Phase 1 and Phase 2 are the first of a series of pilot projects to advance geographic information and geoprocessing service interoperability for critical infrastructure protection. The sponsors of CIPI-1 Phase 1 include Natural Resources Canada, the USGS, and General Dynamics. CIPI-1 Phase 1 focuses on the use of open standards to establish a common understanding of an emergency situation and coordinate incident response.

- **GeoSpatial One-Stop Transportation Pilot.** This is an initiative supporting the U.S. e-government geospatial one-stop initiative aimed at sharing mapping data and services.

- **Conformance & Interoperability Testing & Evaluation Initiative.** This is an OGC interoperability initiative designed to test and evaluate OpenGIS Specification-based interfaces and products that implement them.

- **Geographic Information for Sustainable Development Initial Capability Pilot.** This is the first of a series of projects to help make geographic information more accessible and useful to decision makers working on sustainable development problems. The GISD-ICP is being funded by the U.S. State Department and the U.S. Agency for International Development. Supporting sponsorship comes from Natural Resources Canada and the Federal Geographic Data Committee (on behalf of the Global Spatial Data Infrastructure).

- **OGC Web Services Initiative.** This program is involved with issues related to the Web Map Server, Catalogs, and Sensor Web.
- **Geographic Objects Feasibility Study.** The GO-1 Feasibility Study is sponsored by the Defense Information Systems Agency (DISA) and the Space and Naval Warfare Systems Command (SPAWAR). It is evaluating the possibility of building a set of standard object definitions for constructing geospatial systems. This study has assessed current OGC Web services technology. It also has examined the feasibility of extending that technology to other distributed computing environments, leveraging a robust modeling approach for interoperability.

Conclusion

A spatial operation may consist of a simple spatial search to find some specific location or involve a more advanced query that seeks out attributes shared among a variety of entities. The true power of GIS, however, only becomes apparent when new data are derived from existing data to make an informed decision. As such, a geographic information system should be thought of as a decision support tool, not a computerized mapping program (CMP).

To make a distinction between a GIS, a CMP, or even a simple spreadsheet program, keep in mind that spatial queries are location dependent, whereas attribute queries are aspatial. For example, a company's listing of active leases may have each block coded according to ownership. From this, an attribute query may discern all company-operated leases from minority holdings without the use of a map. Because such a query can be conducted in a spreadsheet without the use of a GIS, this type of query would be considered a standalone attribute query.

But what if management asked how many exploratory blocks (prospects) lie within 5 mi of each operated lease? The answer to such a question would then involve spatial information about the location of these prospects. The power of GIS is further revealed in several other ways. An analyst might consider joining information about infrastructure from within the leases with the existing infrastructure outside the leases. Or the analyst might need to decide whether expansion would provide economic gain. Thus, the joining of spatial locations with aspatial attributes using GIS network analysis can be integrated to develop new answers from preexisting data.

References

[1] Heywood, I., S. Cornelius, and S. Carver. 1998. *An Introduction to Geographic Information Systems*. New York: Addison Wesley Longman. p. 42.

[2] Mason, R.D. and D.A. Lind. 1996. *Statistical Techniques in Business & Economics*. 9th ed. Boston: McGraw-Hill. p. 14.

[3] Stockburger, D.A. *Introductory Statistics*. On-line Course. Southwest Missouri State University.

[4] Mason and Lind, p. 15.

[5] Ibid. p. 16.

[6] Stockburger.

[7] Mason and Lind, p. 16.

[8] Ibid. p. 97.

[9] Chou, Y.H. 1997. *Exploring Spatial Analysis in Geographic Information Systems*. Santa Fe, NM: OnWord. p. 141.

[10] Mitchell, A. 1999. *The ESRI Guide to GIS Analysis*. Redlands, CA: ESRI. p. 51.

[11] Ibid.

[12] Ibid. p. 50.

[13] Ibid.

[14] Berry, J.K. 1997. *Spatial Reasoning for Effective GIS*. Fort Collins, CO: GIS World. p. 26.

[15] Mason and Lind, p. 95.

[16] Mitchell, p. 48.

[17] Ibid. p. 47.

[18] Mason and Lind, pp. 347–350.

[19] Coburn, T.C. and J.M. Yarus, eds. 2000. "AAPG Computer Applications in Geology, No. 4." *Geographic Information Systems in Petroleum Exploration and Development*. Tulsa: American Association of Petroleum Geologists. p. 296.

[20] Ibid.

[21] Chou, p. 187.

[22] Ibid. p. 189.

[23] Mitchell, p. 44.

[24] Chou, p. 192.

[25] Ibid. p. 193.

26 Jones, C. 1997. *Geographical Information Systems and Computer Cartography*. Essex, England: Longman. p. 29.

27 Burrough, P.A. and R.A. McDonnell. 1998. *Principles of Geographic Information Systems*. Oxford: Oxford University Press. p. 58.

28 Foresman, T.W. ed. 1998. *The History of Geographic Information Systems: Perspectives from the Pioneers*. Upper Saddle River, NJ: Prentice Hall. p. 60.

29 Berry, p. 123.

30 Burrough and McDonnell, pp. 52–57.

31 Chou, p. 67.

32 Berry, p. 107.

33 Burrough and McDonnell, p. 58.

34 Zeiler, M. 1999. *Modeling Our World—The ESRI Guide to Geodatabase Design*. Redlands, CA: ESRI. p. 128.

35 Korte, G.B. 1997. *The GIS Book—Understanding the Value and Implementations of Geographic Information Systems*. 4th ed. Santa Fe, NM: OnWord. p. 115.

36 DeMers, M.N. 1997. *Fundamentals of Geographic Information Systems*. New York: Wiley & Sons. p. 111.

37 Heywood, et al. p. 53.

38 Ibid. p. 55.

39 Foresman, p. 53.

40 Heywood, et al. pp. 59, 123.

41 Ibid. p. 59.

42 Ibid. p. 60.

43 Ibid. p. 123.

44 Chou, p. 234.

45 Ibid. p. 230.

46 Heywood, et al. p. 124.

47 Ibid.

48 Chou, p. 153.

49 Ibid. p. 154.

50 Zeiler, p. 112.

51 Jones, p. 171.

52 Heywood, et al. p. 112.

53 Burrough and McDonnell, p. 166.

54 Heywood, et al. p. 111.

[55] *Using ArcView GIS*. 1996. Redlands, CA: Environmental Systems Research Institute (ESRI). p. 244.

[56] Chou, p. 156.

[57] Ibid. p. 157.

[58] DeMers, p. 246.

[59] Zeiler, p. 162.

[60] Chou, p. 318.

[61] Burrough and McDonnell, p. 64.

[62] Zeiler, p. 164.

[63] Chou, p. 322.

[64] Ibid. p. 324.

[65] Ibid. p. 334.

[66] Ibid. p. 267.

[67] Ibid.

[68] Ibid. p. 268.

[69] Ibid. p. 269.

[70] Heywood, et al. pp. 71–72.

[71] Laudon, K.C. and J.P. Laudon. 1996. *Management Information Systems—Organization and Technology*. 4th ed. Upper Saddle River, NJ: Prentice-Hall. p. 280.

[72] Heywood, et al. p. 73.

[73] Ibid.

[74] Zeiler, p. 16.

[75] von Meyer, N.R. and R.S. Oppman, eds. 1999. *Enterprise GIS*. Urban and Regional Information Systems Association (URISA). p. 10.

4

Geodesy

In chapter 3, we summarized the concepts of GIS modeling and how spatial relationships can be defined and interpreted by location, attributes, and topology. Yet before we can perform any type of analysis, we must make sure all entities within the database are precisely defined with respect to a common frame of reference or coordinate system. Otherwise, we will be working with disparate data sets that are not geographically related.

If we do not ensure correct and consistent geodetic positioning, resource delineation and oilfield operations can be seriously affected. A geologist, for example, may base all work on NAD27 (North American Datum of 1927) and forget to convert prospect coordinates to WGS84 (World Geodetic System of 1984). If the field superintendent then uses a GPS receiver to spot the well, drilling crews will rig up equipment on the wrong location (see Fig. 4–1).

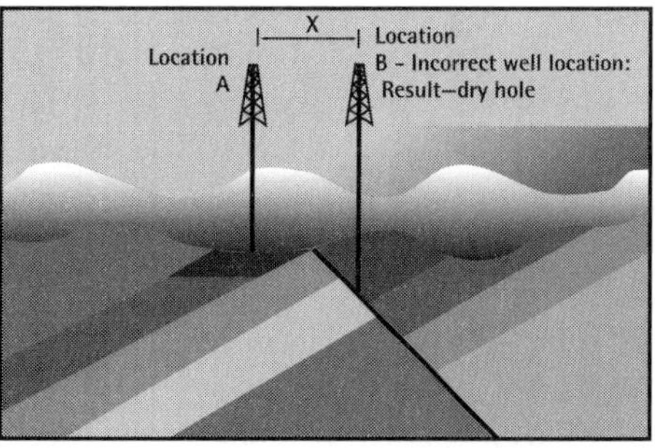

Fig. 4–1 Importance of Using Same Geodetic Parameters (source: OGJ[1] and Martin Rayson, Quest, Ltd.)

Such mistakes may even cross into the political arena. In some cases, neighboring countries might use different systems to locate boundaries [see Fig. 4–2 (Plate 4–1)]. The end result in this case may be lengthy delays in field development and even international disputes.

With continued advances being made in computer-based mapping and the widespread use of GPS, an entirely new set of problems now faces GIS users. In fact, simple mistakes can lead to dry holes, cases of litigation, and angry landowners. These mistakes can result from something as simple as applying the wrong sign convention in software conversion or disagreeing over geodetic parameters. Errors also can occur from falsely believing that GPS receivers base elevation on mean sea level.

This is where geodesy comes into play. In simple terms, geodesy forms the backbone of our GIS.

Definition

Webster defines geodesy as "a branch of applied mathematics concerned with the determination of the size and shape of the earth and the exact positions of points on its surface and with the description of variations of its gravity field."[2] In practice, geodesy uses the principles of astronomy, geophysics, and geology and applies these within the capabilities of modern surveying and engineering technologies. As users of GIS, we totally depend upon geospatial exactness. Thus we must make sure all data-gathering instruments, survey methodologies, and mathematical conversions are accurately employed each and every time.

> *In geodesy, we aim to understand how to describe the position of points on, above, or below the Earth's surface. To do this a geodetic datum must be used and the accuracy to which a point can be described depends upon the care taken in defining this system.*
>
> —Martin Rayson

Fundamentals

The study of geodesy involves six primary topics:

1. Reference surfaces (topography, geoid, and ellipsoid)
2. Earth coordinate systems, both geographic (latitudes and longitudes) and Cartesian *(X,Y)*
3. Datums
4. Datum transformations
5. Projections (see chapter 5)
6. Grids

Each category adds pertinent information to our picture of the world, much like pieces of a jigsaw puzzle. As the ultimate goal is to measure real-world features accurately, we must repeatedly and precisely measure features on and within the Earth so that we can perform geospatial computations.

Reference Surfaces

The tools used to illustrate the Earth's shape and size consist of *topographical*, *geoidal*, and *ellipsoidal* surfaces (see Fig. 4–3).

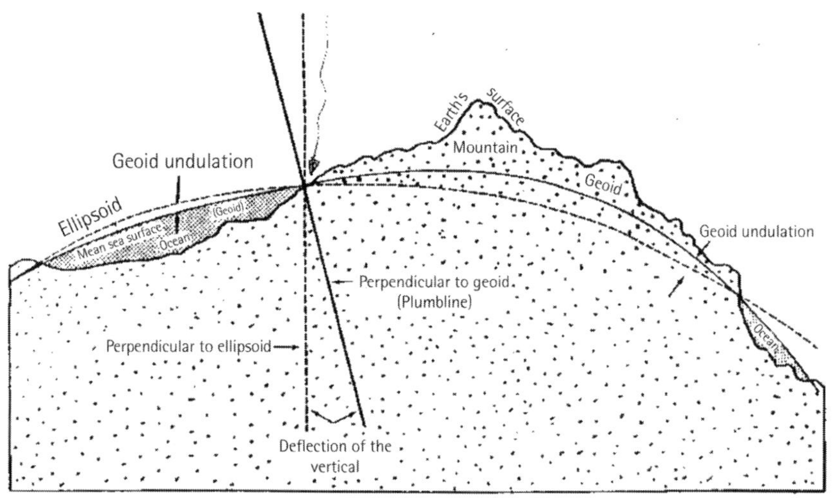

Fig. 4–3 The Three Surfaces: The Earth, Geoid, and Ellipsoid

Our discussion begins with the topographical (terrestrial) surface as it forms our basis for physical reality. This is the place on which we conduct most of our observations.

The Earth is not smooth. Instead, it contains canyons, rivers, mountains, and valleys. And it is constantly in motion, with earthquakes, seafloor spreading, and subduction zones continuously destroying, adding, and recycling material. In essence, the Earth has an extremely complicated shape that is impossible to describe mathematically. In order to geospatially model and analyze this surface with higher levels of precision, we must turn to the equipotential surface of the Earth's gravity field known as the geoid.

The geoid

"The geoid is a surface along which the gravity potential is everywhere equal and to which the direction of gravity is perpendicular," says Martin Rayson, a geodesist with Quality Engineering and Survey Technology Ltd. (Quest), Houston. "You can visualize this surface by looking at how mean sea level, absent from the effects of tides, would look around the world."[3]

Basically, the geoidal surface provides geodesists with a go-between from the terrestrial surface to the ultimate computational surface, the ellipsoid. Yet the Earth's shape and mass are affected by variations in topography, density anomalies, rotational effects, distribution of water bodies, and celestial interaction. Consequently, the geoid, like the terrestrial surface, also consists of an irregular, albeit smoother shape. Thus, it cannot be easily described mathematically, again discounting its use as a computational surface. (By some accounts, just trying to represent the geoid would require a mathematical series with terms to the 18th power.[4])

To define this equipotential surface, geodesists apply a variety of techniques. Onshore measurements, for example, depend mostly on relative gravity data. (The National Oceanic and Atmospheric Association's website at http://www.ngs.noaa.gov/TOOLS/ contains a program that computes values of Earth's magnetic field, including declination.) This data is collected with gravimeters, which are sensitive, spring-mounted weights. The effects of subsurface bodies, with varying degrees of size and density, obviously affect such readings.

Conversely, offshore measurements rely mostly upon continuous measurements of sea level readings taken around the world and averaged for a period of 18.6 years.[5] Obviously, the effects of the Moon and Sun on the oceans' tides, along with the Coriolis effect, will affect sea level readings around the world. As a result of these surface, subsurface, and celestial anomalies, the geoidal surface varies from ocean to ocean and from continent to continent.

Nevertheless, it is possible to tie these oceanic and continental measurements together. The result is that a fairly accurate equipotential surface over the surface of the world has been generated that continues to be refined with additional information (see Box 4–1).

Box 4–1
Gravitation and Potential

An examination of the concepts of gravitation and potential helps us to understand equipotential surfaces.

Gravitation is the universal effect resulting in all bodies attracting one another as summed up in Newton's Law of Universal Gravitation. Essentially, the Earth has a gravity field that attracts all objects to the center of the field (Quest, 1997).

As the shape of the Earth is an ellipsoid, it follows that certain parts of its surface will lie farther from the center of mass than other parts. As such, the equatorial regions have been found to be farther from the center of mass than the polar regions. Such variations produce different values of gravity at different latitudes with slightly weaker gravitational forces occurring in the equatorial regions.

Potential is the measure of gravitational force and is defined as the amount of energy required to move an element of unit mass from infinity to a given point–against the gravity field. It is calculated with the simple expression:

Potential = Force x Distance

The SI unit of force is called the Newton, and potential is given the unit Newton-meters (Nm). Thus, if 10 N are required to compress a spring 3.5 m, the spring will have a potential of 35 Nm. Therefore, because gravity is a force, and if a mass is lifted against this force, the mass will gain potential energy. "When you lift a 100-kg weight off the floor, your arms and legs will tell you that you have given the weight a good deal of potential energy, where the amount of energy it possesses could be confirmed by dropping the weight on your foot," Rayson said. "A 200-kg weight dropped from 1 m, however, will hurt your toes every bit as much as a 100-kg weight dropped from 2 m as both have the same potential energy."

Now that *gravitation* and *potential* have been described, it is easier to picture the geoidal surface. "Imagine lifting a unit mass from the center of the Earth to somewhere near the Earth's surface," he said. "By doing so, the unit mass gains a certain potential." Next, take another unit mass and lift it to another point so it has exactly the same potential as the first. Repeat this exercise until an undulating "blanket" of unit masses encompasses the Earth. "This blanket, which is called an *equipotential* surface, then becomes a surface on which the potential at every point is identical." It should be noted that an infinite number of equipotential surfaces do exist above or below the Earth's surface. However, the equipotential surface that most concerns our discussion is the geoid, which equates to mean sea level. The geoid thus became the surface to which all vertical coordinates (on land) were referenced for contour charts and maps.

Ellipsoids

Although the geoid is much smoother than the terrestrial surface, this surface still contains too many undulations to produce a workable model for computations. Thus, the geoid cannot be used for geospatial calculations on a horizontal plane. (This is not true for measurements in the vertical plane, to be discussed in chapter 5.)

Fortunately, astronomic measurements on the Earth's surface, gravity observations, and the tracking of orbiting satellites show that the geoid closely conforms to an ellipsoid of revolution.

(In geodesy, the terms *ellipsoid* and *spheroid* are used interchangeably.) "Ellipsoids are idealized surfaces and do not exist," Rayson says. "But because they have an exact mathematical definition, we can model the geoid and accurately define positions on the surface."[6] For a global ellipsoid, the separation between it and the geoidal surface is a maximum of 100 m for any point on the Earth. On the other hand, variations between the ellipsoid and Earth's surface can be quite large.

The geoid and ellipsoid therefore have a special relationship, as the latter is used to model the former.

First, imagine the geoid with all its undulations (see Fig. 4–4). Next, take a best-fit surface and place it within the geoid so that all the wrinkles and deformities, produced by variations in the Earth's physical makeup, are removed. If this is performed correctly, we end up with a best-fit surface that produces a smooth ellipsoidal shape that can be mathematically manipulated.

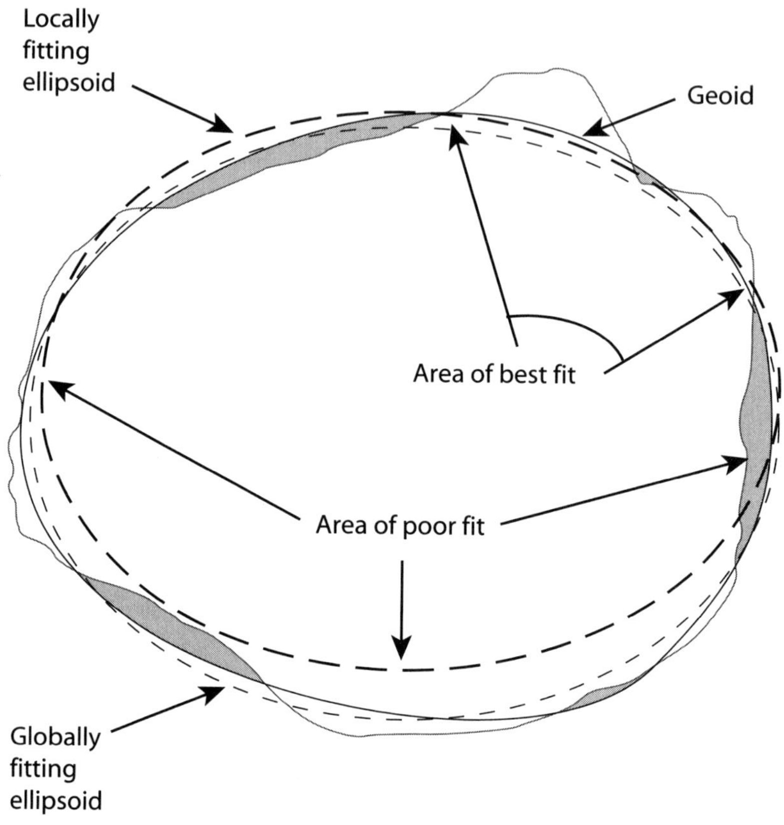

Fig. 4–4 Different Models of the Earth
(source: courtesy of U.K. Offshore Operators Association)

Although four parameters are used to define an ellipsoid, it only takes two dimensions to describe an ellipsoid. The first is the semimajor axis, *a*, represented by the radius of the Earth taken at the equator. The second is flattening, *f*, which indicates how closely an ellipsoid approaches a spherical shape (see Fig. 4–5).

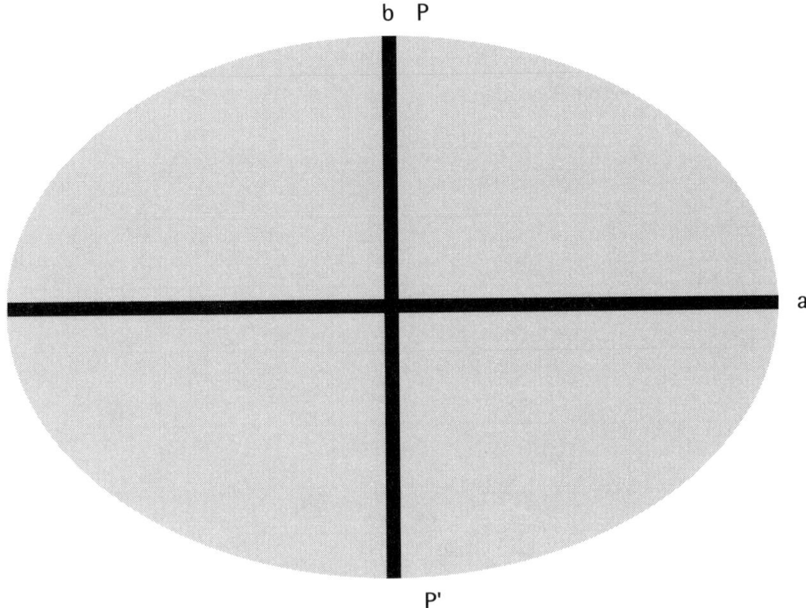

a = Semimajor axis
b = Semiminor axis
f = Flattening = (a-b)/a
PP' = Ellipsoid axis of revolution

Fig. 4–5 Elements of an Ellipse

The Earth is in fact flattened slightly at the poles (by about 1 part in 300) but bulges somewhat at the equator. (For a perfect circle, flattening is equal to 0.) Consequently, the geometrical figure used to most nearly approximate the shape of the Earth can be described through an oblate ellipsoid of revolution. This is rotated about its semiminor axis (polar axis), *b*.

By convention, geodesists use the semimajor axis and flattening to uniquely define the reference ellipsoid, where:

$$f = (a - b)/a$$

The most common reference ellipsoids are shown in Table 4–1.

Datum	Usage	Ellipsoid	dX	dY	dZ
Ain el Abd 1970	Bahrain	International 1924	-150	-250	-1
Ain el Abd 1970	Saudi Arabia	International 1924	-143	-236	7
Australian Geodetic 1984	Australia	Australian National	-134	-48	149
Bogota Observatory	Colombia	International 1924	307	304	-318
Bukit Rimpah	Indonesia (Bangka & Belitung Ids)	Bessel 1841	-384	664	-48
Campo Inchauspe	Argentina	International 1924	-148	136	90
Corrego Alegre	Brazil	International 1924	-206	172	-6
European 1950	Egypt	International 1924	-130	-117	-151
Gunung Segara	Indonesia (Kalimantan)	Bessel 1841	-403	684	41
Indian	Bangladesh	Everest (India 1830)	282	726	254
Indian	Pakistan	Everest (Pakistan)	283	682	231
Indonesian 1974	Indonesia	Indonesian 1974	-24	-15	5
Luzon	Philippines (Excluding Mindanao)	Clarke 1866	-133	-77	-51
Nahrwan	Saudi Arabia	Clarke 1880	-243	-192	477
Nahrwan	United Arab Emirates	Clarke 1880	-249	-156	381
Naparima BWI	Trinidad & Tobago	International 1924	-10	375	165
NAD 1927	Alaska (Excluding Aleutian Ids)	Clarke 1866	-5	135	172
NAD 1927	Canada (Alberta; British Columbia)	Clarke 1866	-7	162	188
NAD 1927	Canada (Manitoba; Ontario)	Clarke 1866	-9	157	184
NAD 1927	Canada (New Brunswick, Newfoundland,Nova Scotia,Quebec)	Clarke 1866	-22	160	190
NAD 1927	MEAN FOR CONUS	Clarke 1866	-8	160	176
NAD 1927	MEAN East of Mississippi	Clarke 1866	-9	161	179
NAD 1927	MEAN West of Mississippi River	Clarke 1866	-8	159	175
NAD 1927	Mexico	Clarke 1866	-12	130	190
NAD 1983	Alaska (Excluding Aleutian Ids)	GRS 80	0	0	0
NAD 1983	Canada	GRS 80	0	0	0
NAD 1983	CONUS	GRS 80	0	0	0
North Sahara 1959	Algeria	Clarke 1880	-186	-93	310
Old Egyptian 1907	Egypt	Helmert 1906	-130	110	-13
South American 1969	Trinidad & Tobago	South American 1969	-45	12	-33
South American 1969	Venezuela	South American 1969	-45	8	-33
South Asia	Singapore	Modified Fischer 1960	7	-10	-26
Voirol 1960	Algeria	Clarke 1880	-123	-206	219
WGS 1972	Global Definition	WGS 72	0	0	0
WGS 1984	Global Definition	WGS 84	0	0	0

Table 4–1 A Few Datums
(source: Quality Engineering and Survey Technology Ltd.)

It is important to note that traditional vertical measurements (spirit and trigonometric leveling) and contour lines on the Earth are reported relative to the geoid, not the ellipsoid (see Fig. 4–6).

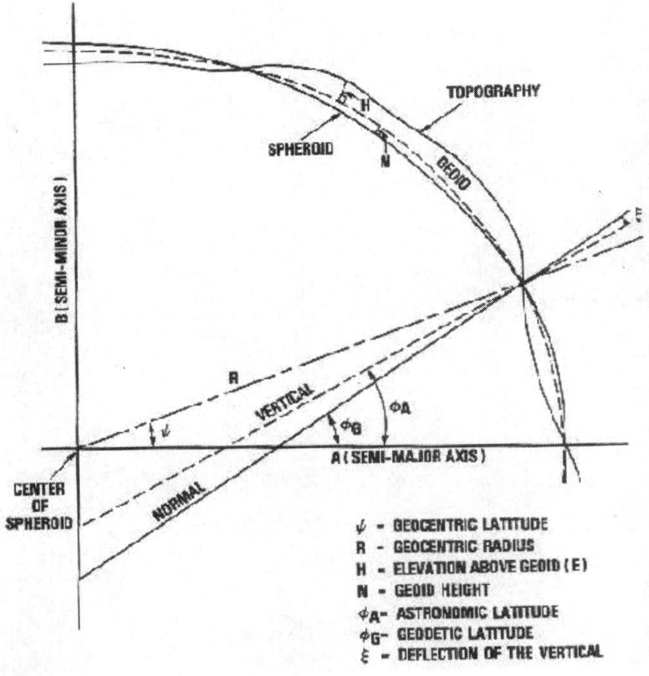

Fig. 4–6 Relationships among Geodetic Surfaces

Conversely, elevation readings taken with a GPS receiver are uniquely determined with respect to the ellipsoid (see chapter 6).

Coordinates

Once a reference ellipsoid has been chosen, the means for pinpointing features can be conducted through measurements of angular or planar Cartesian coordinates. With angular coordinates, positions in 3-D space can be expressed by two angles, latitude and longitude, and ellipsoidal height (see Fig. 4–7).

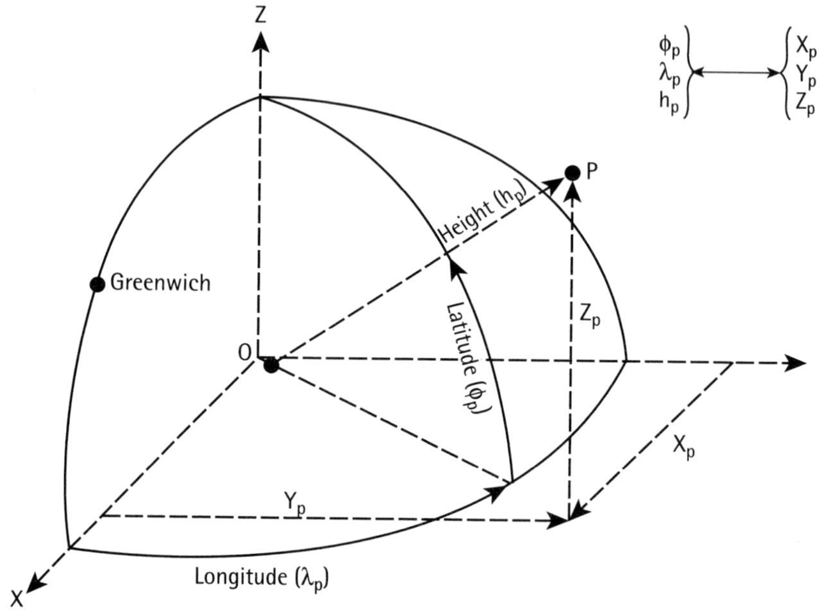

Fig. 4–7 Terrestrial Coordinate Systems

Latitude, f, is measured with respect to the equatorial plane using positive values above the equator and negative below. Longitude, l, is measured with respect to a reference meridian plane established at the Greenwich Observatory in England. Positive longitude values are noted in the Eastern Hemisphere and negative values in the Western Hemisphere. It can be seen that longitude is arbitrarily defined, anchored by international agreement. In contrast, latitude is a natural coordinate, as related by the Earth's rotation on its axis.

Coordinates can be defined either through astronomical or geodetic methods (see Box 4–2). A sailor, for example, can determine latitude, longitude, and azimuth with a telescope, timepiece, and star almanac. On the other hand, a surveyor may use GPS-derived readings to map latitude, longitude, ellipsoidal height, azimuth, and velocity in a geodetic system.

Box 4–2

Astronomic Coordinates

1. **Astronomical latitude.** The angle that the normal to the geoid (plumb line) makes with the equatorial plane. Ancient mariners used this angle to define their position at sea. The calculation of latitude has never been a problem as the North Star (Polaris) provides a fixed point of measurement to determine this coordinate. In the Southern Hemisphere, the Southern Cross, with adjustments, is used.
2. **Astronomical longitude.** The angle between two planes that both contain the axis of rotation of the Earth. One plane is parallel to the vertical at a given point and the other is parallel to some arbitrary defined reference plane (e.g., the Greenwich Meridian). Ancient mariners were unable to calculate longitude until the development of reliable time pieces in the mid-1770s (sea clock invented by John Harrison).
3. **Astronomical azimuth.** The true bearing between two points *P* and *P1*, both of which lie on the surface of the Earth. One point contains the vertical at P and the astronomical North Pole, while the other contains the verticals of the two points.

Geodetic Coordinates

1. **Geodetic or geographic longitude.** The angular distance measured on the terrestrial equator from the intersection of a fixed meridian and the equator to the foot of the meridian through the observer's position. The meridian through Greenwich, England, is taken as the fixed meridian.
2. Geodetic or geographic latitude. The angle that the normal to the ellipsoid makes with the equatorial plane.
3. Deflection of the vertical. The angle between the normal to the ellipsoid and the normal to the geoid at a point.
4. Geocentric latitude. The angle the radius vector (from the ellipsoid center to the point) makes with the equatorial plane.

Sources: Morgan, p. 28; Quest CD ROM.

Cartesian coordinates

Unfortunately, although angular units such as latitudes and longitudes may be fine for defining precise locations, they are not always easy to work with mathematically. For example, think about measuring distances on a globe with a straight edge, or measuring angles on a curved surface with a protractor. Another problem concerns the meridians and parallels. These may well be perpendicular to each other at the point of intersection, but the unit distance along a meridian will change in relation to the parallel as one travels farther from this point. In other words, the distance of 1° longitude is not always equal to 1° latitude (see Table 4–2).[7]

Length of one degree of longitude:	
GRS 80 Ellipsoid	
Latitude	**Kilometers**
0°	111.32
10°	109.64
20°	104.65
30°	96.49
40°	85.39
50°	71.70
60°	55.80
70°	38.19
80°	19.39
90°	0.00

Table 4–2 GRS80 Ellipsoid

The Cartesian X-Y and X-Y-Z coordinate systems, on the other hand, provide an efficient means for measuring distance and direction (see chapter 5). This is especially true for small areas where the curvature of the Earth is not pronounced. With the Cartesian coordinate system, the X-axis and Y-axis remain perpendicular to each other, with identical unit distances in either direction. Furthermore, when a third dimension, Z, is added, these properties still hold.

The Datum Connection

Having discussed ellipsoids and geoids in relation to the Earth's surface, we can return to defining coordinates in a system that can be used within a GIS. As previously stated, modeling the physical Earth, with all of its real-life features, is not practical from a mathematical point of view. Yet we must find a means to precisely define points, lines, areas, surfaces, and volumes, or it becomes impossible to perform geospatial analyses or make a useable map. But even the ellipsoid, a mathematical best-fit of the geoid, falls short of providing a complete tool for geospatial work. This is where the *datum connection* comes into play.

"In simplest terms, a geodetic datum consists of a reference ellipsoid that is fixed in some manner to the physical Earth."[8] A datum, in other words, embodies a reference ellipsoid that is tied to a coordinate frame with a fixed origin. There are two types of datums: *global* and *astrogeodetic (local)*.

Global datums

A global datum depends on the best-fit selection of an ellipsoid for the entire Earth. Global datums fit the reference ellipsoid to the center of mass of the Earth through the coincident alignment of the semimajor axis with the Earth's mean spin axis.[9] These datums are Earth centered and Earth fixed to the best approximation of the geoid.

The best-known global datum is the WGS84. This is a coordinate system developed through:

- Numerous Doppler points (1591)
- Satellite laser ranging (SLR)
- Very long baseline interferometry (VLBI)
- A classified geoid developed from surface gravity and geoid heights

WGS84 evolved through developments from prior WGS systems: WGS60, WGS66, and WGS72. Each development improved accuracy as satellite data continuously provided more detail of the Earth's shape. (It should be noted that the naming convention for the WGS series stopped with WGS84, although updates continue even today.)

The Navstar GPS, to be discussed in chapter 6, uses the WGS84 datum. The original accuracy of the WGS84 datum was on the order of 1–2 m. Today, absolute accuracy of WGS84 is about 20–30 cm.

Local datums

The best way to imagine a local datum is to consider two globes, one representing a geoid and the other an ellipsoid. Place one inside the other so that they are tangentially pressed against each other. The exact point where the inner globe touches the outer globe will define the origin point. In Figure 4–8, the perpendicular to the ellipsoid is made coincident to the geoid at the datum origin.

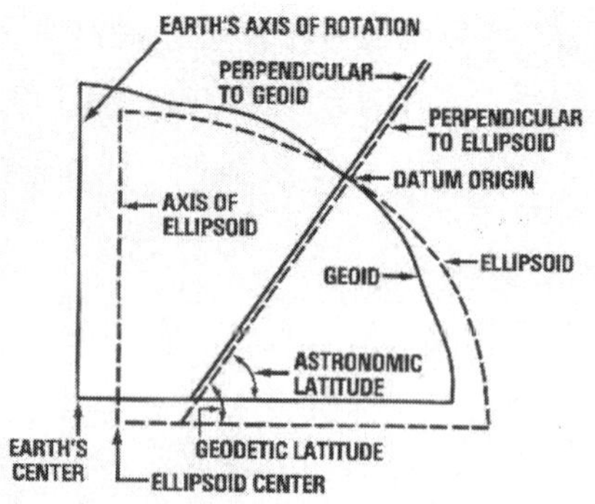

Fig. 4–8 A Single Astronomical Station Datum Orientation

Local datums are used for specific regions such as North America, Europe, Africa, Asia, and Europe (see Fig. 4–9).

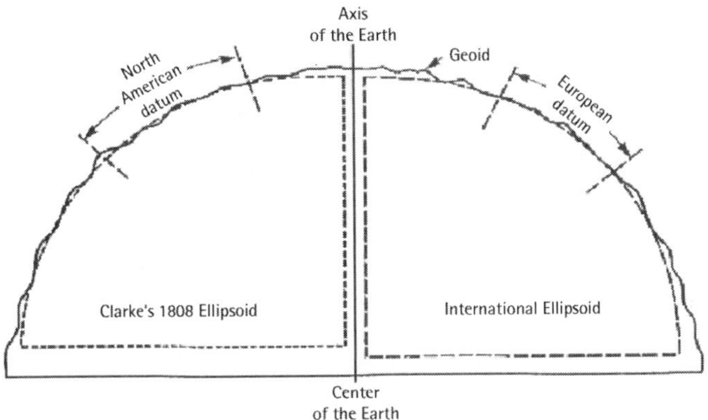

Fig. 4–9 Regional Datums Are Local Datums

In such cases, the coordinate frame is determined and an ellipsoid is individually defined to minimize the local geoid-ellipsoid separation. There are literally hundreds of local datums around the world that bear no relation to one another (see Table 4–1 and Fig. 4–10).

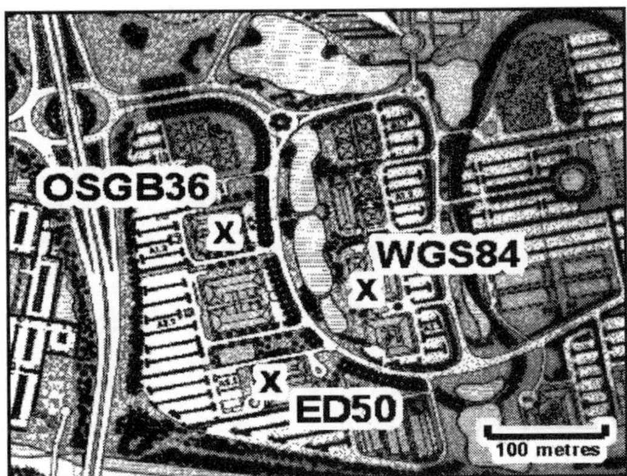

Fig. 4–10 Three Different Datums, Three Different Locations on the Earth's Surface (source: U.K. Offshore Operations Association)[10]

A well-known example includes the North American Datum of 1927 (NAD27), which uses the Clarke 1866 Reference Ellipsoid and a coordinate frame tied to Meade's Ranch, Kansas.

The position and orientation of a local coordinate frame can be defined in one of two ways: explicitly through the origin or through astronomical means.

At the point of origin, where astronomic latitude and longitude have been determined, regional and local datums can be defined using seven parameters (see Fig. 4–6):

- a, semimajor axis of the reference ellipsoid
- f, flattening of the reference ellipsoid
- ξ, component of the deflection of the vertical in the meridian at the datum origin
- η, component of the deflection of the vertical in the prime vertical at the datum origin
- α, geodetic azimuth from the origin at another point in the system
- N, geoid height at the datum origin, i.e., the difference between the ellipsoid and the geoid at the origin
- The condition that the ellipsoid semiminor axis and the Earth's mean rotation axis be parallel for global datums[11]

For local datums, however, the rotational parameters mentioned above must be used in the datum transformation model.

In the simplest form, the local datum uses a single astronomic position as the point of origin. The ellipsoid and geoid are assumed to be coincident and tangential at the origin. Thus, $N = \xi = \eta = 0$. A datum origin with only a single astronomic point, however, may produce large geoid-ellipsoid separations over the extent of the datum. The farther you travel from the origin, the greater the inaccuracies may become in relation to how the ellipsoid describes the geoid. This is because the ellipsoid is not Earth-centered. Its rotational axis is parallel, but not coincident, with the mean rotational axis of the Earth.[12]

A slightly more complex datum orientation, known as the astrogeodetic datum, can be used with greater accuracy over larger areas. "If numerous points over the local area have known astronomical and geodetic coordinates, it is possible to calculate the deviation of the vertical at all these points so that the geoid-ellipsoid separation can be defined," Rayson said.[13] Continental areas that encompass Europe, Southeast Asia, Russia, North America, South America, India, South Africa, and Australia have regionally defined astrogeodetic datums. Individual countries within these regions fix their own single-point astronomical datums.

GIS users should be acutely aware that positions derived from different astronomically oriented datums cannot be directly related to each other in geodetic computations.

Local datums may someday become a thing of the past as satellite navigation rapidly becomes the tool of choice. However, because of the immense amount of historical data still referenced to the local datum, it is highly unlikely they will be phased out soon.

Datum transformation

"The datum transformation, or the mathematical conversion of geographic coordinates from one geodetic datum to another, is currently the largest cause of error," said Rayson (see Fig. 4–11).[14] To overcome this, the coordinates used to describe a position must be linked intrinsically to a geodetic datum, which in turn has a specific reference frame.

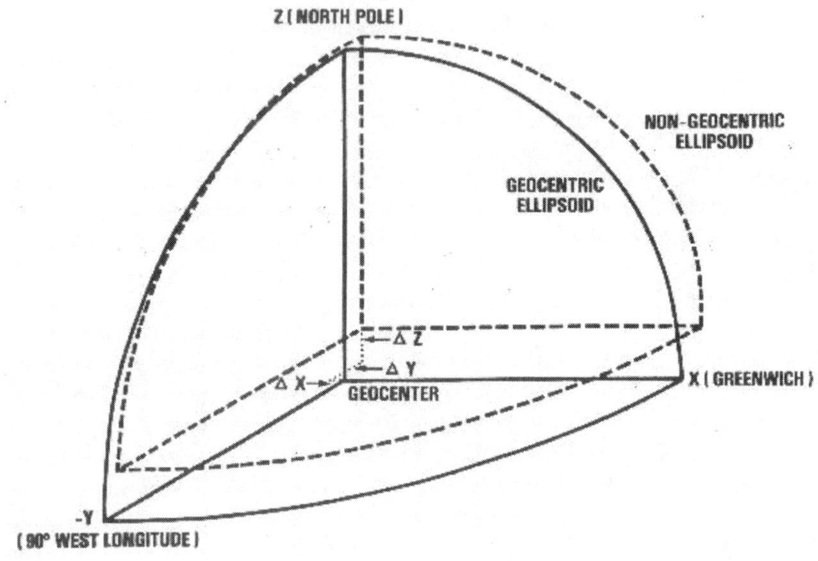

Fig. 4–11 Geometrical View of a Datum Transformation

These two are combined to create global or local datums as described above. For coordinates to have any physical meaning, the datum must always be listed for precise reference. Thus, the datum serves as an anchor for the coordinate framework. Excluding this information will leave the user stranded in terms of position.[15]

When given coordinates of latitude and longitude, for example, we may mistakenly believe that we have the precise position. Yet nothing can be farther from the truth.

In fact, a physical point on the face of the Earth may be defined by a multitude of latitude and longitude values, depending on the choice of datum. Simply stating that a compressor is located at 38° 39' 27" N and 107° 44' 22" W does not provide the exact location. It can vary by meters to kilometers in different systems, depending on which ellipsoid and datum are used. Thus, *latitude and longitude values are not unique*.

"This oversight is often neglected and causes confusion when two or more data sets of different vintages are combined," Rayson says. It "is especially troublesome for companies that have foreign operations."[16] For example, when working with the local country, it is standard policy to provide maps in the local datum and coordinate system.

However, to build a multinational data set in GIS, there are different concerns. According to Burrough, "It is important to ensure cross-border equivalence of ellipsoids, projections, and base levels."[17] And just as important, it is imperative that all calculations used in the transformation be double-checked by qualified personnel.

Rayson, for example, encountered a situation in the Caribbean where a drilling location was identified using a 3-D seismic data volume. This was derived in the WGS84 datum.[18] In this case, local workers and government agencies required a working system in the Old Trinidad 1903 (OTD1903) datum to maintain parity with prior work.

The geodetic coordinates of the location, as given in WGS84, were:

60° 35' 26" W

10° 20' 45" N

Converting to OTD1903, however, provided the following coordinates:

60° 35' 28.86" W

10° 29' 31.64" N

At this point, note the numeric difference between the two longitudes and latitudes. They are supposed to pinpoint the same location on the physical Earth. Because they are tied to different datums, they produce different values of latitude and longitude.[19] In order to unify these two coordinates into one system, a variety of datum transformation methods can be utilized, depending on ease of use and the availability of data (see Box 4–3).

Box 4–3
Datum Transformation Methods

Method	Advantages	Disadvantages
3-Parameter	• Easy; widely used • Liner; shifts additive	• Needs localized DX, DY, DZ for high accuracy
5-Parameter	• Typically more accurate than 3-Parameter • Easy	• Parameters not widely available
7-Parameter	• Still more accurate than above	• Parameters not widely available and still needs many known points to derive parameters that give high accuracy
Multiple regression	• More accurate than widely used 3-Parameter shifts • Easy	• Parameters available for very few areas
Gridded surface fit	• Fast • Very accurate • Covers entire area • Continuous • Consistent	• Available in few areas but can be best method when available
Grid-to-grid	• Fast • Can be very accurate	• Invitation for trouble • Best left to an expert

Source: Jim Morgan.
Note: The methodology for executing these transformation shifts are beyond the scope of this book.

Common Mistakes

To show how easy it is to misplace oilfield assets, we look at two of the many ways in which we can produce computational errors.

An incorrect sign change

The first case describes a situation in which a sign change within the seven-parameter transformation procedure produced a locational error of more than 1 km. The seven-parameter transformation, one of the more accurate geodetic models in use today, includes:

- *Three translation parameters* that define the difference between the spheroidal centers of Datum A and Datum B. These are expressed by ΔX, ΔY, and ΔZ (in m), within the Cartesian coordinate frame.
- *Three rotation parameters*, where rotations about the X, Y, and Z axes serve to align the axes of the spheroids (seconds of arc). Two principle rotation conventions are used. The first is the Bursa-Wolfe model, which has a positive clockwise convention. The second is the coordinate frame rotation, which has a positive counterclockwise convention.
- *One scale parameter* that matches the size of the two spheroids (ppm).

As will be shown, when dealing with the three translation parameters, it is especially important to make sure that the correct sign is applied. Otherwise, large errors will result. For example, to go from WGS84 to OTD1903, the parameters may be:

$$\Delta X = +\ 61.00 \text{ m}$$

$$\Delta Y = -\ 285.00 \text{ m}$$

$$\Delta Z = -\ 471.00 \text{ m}$$

On the other hand, from OTD1903 to WGS84, the magnitude of the translation parameters remains the same, but the signs must be reversed.

To perform a seven-parameter datum transformation, Rayson says a four-step procedure should be used.

1. For the WGS84 datum, convert the geodetic coordinate using WGS84 reference ellipsoid parameters to their equivalent Cartesian coordinates.
2. Enter the Cartesian coordinates (Datum 1) along with the given transformation parameters into datum transformation Equation 1 (see Fig. 4–12). A positive clockwise convention is assumed in this example.
3. Compute the Cartesian coordinates (Datum 2) of the position in the OTD1903 datum.
4. Convert the Cartesian coordinates to their equivalent geodetic coordinates using the OTD1903 reference ellipsoid parameters.

$$\begin{bmatrix} X \\ Y \\ Z \end{bmatrix}_2 = (1 + s{*}10^{-6}){*} \begin{bmatrix} 1 & -\theta_Z & +\theta_Y \\ +\theta_Z & 1 & -\theta_X \\ -\theta_Y & -\theta_X & 1 \end{bmatrix} \begin{bmatrix} X \\ Y \\ Z \end{bmatrix}_1 + \begin{bmatrix} \Delta X \\ \Delta Y \\ \Delta Z \end{bmatrix}$$

7-Parameter datum transformation

$$\begin{bmatrix} X \\ Y \\ Z \end{bmatrix}_2 = \begin{bmatrix} X \\ Y \\ Z \end{bmatrix}_1 + \begin{bmatrix} \Delta X \\ \Delta Y \\ \Delta Z \end{bmatrix}$$

3-Parameter datum transformation (can be used when changes in rotation parameters will not affect answer

Fig. 4–12 Datum Transformation Equations

To understand how important it is to check all parameters, notice what happens when the signs are not correctly applied. For example, if the OTD1903 to WGS84 sign convention is used, instead of the opposite, we end up with:

60° 35' 23.14" W

10° 30' 05.12" N

As compared to the correct answer:

60° 35' 28.86" W

10° 29' 31.64" N

We find that the well location would have been off by more than 1 km. Obvious repercussions to a mistake of this order might be a dry hole or one that may have been drilled outside a lease boundary (see Fig. 4–1).

Effects on the bottom line

For the second example, we look at the effect two different transformation methods have on volumetric reserves. A basic conversion problem found in the Gulf of Mexico (GOM) consists of converting WGS84 GPS data into NAD27 data. This is a typical situation as most oil companies use the NAD27 datum to maintain a match with their historical databases. According to Pete Worsnop, a geodesist with Quest, there are two ways to perform this calculation:

1. Make use of a fixed three-parameter shift (Commercial software packages like Blue Marble's Geographical Calculator have this function built into the software.)
2. Use the published government tables of the continental U.S. (CONUS), which calculates a variable shift for every 7.5 min. of latitude and longitude.

Table 4–3 shows the calculations made by Blue Marble and CONUS, utilizing both geographic and UTM grid values.

WGS 84 Datum		**NAD 27 Datum**	
1. 3-Parameter Shift			
26° 30' 00.000 N	92° 00' 00.000 W	26° 29' 58.729 N	91° 59' 59.510 W
2931 445.66 N	599 653.28 E	2931 241.72 N	599 668.99 E
2. CONUS			
26° 30' 00.000 N	92° 00' 00.000 W	26° 29' 58.937 N	91° 59' 59.781 W
2931 445.66 N	599 653.28 E	2931 248.06 N	599 661.42 E

Table 4–3 Gulf of Mexico Datum Conversion

As shown, the difference between the modern software package and the manually derived tables can be seemingly small. It resulted in a shift of only 9.87 m at 310°.

"Yet," says Worsnop, "if we look at the effect this shift can have on an equity calculation made at a block boundary between two oil companies, with an oilfield straddling both sides, we can see that a lot is at stake."[20] For example, if one company insists on using the software, and the other the CONUS tables, the 9.87 m difference can result in a huge difference in volumetric reserves. This depends on the length of the strip under contention, the true vertical thickness of the pay zones, average porosity, average water saturation, and recovery factor:

$$\text{Producible Hydrocarbon Volume} = A \text{ x } H \text{ x } \mathbf{\Phi} \text{ x } (1 - S_w) \text{ x } R_f$$

where

A is the areal size of the field,

H is the aggregate true vertical pay-zone thickness,

$\mathbf{\Phi}$ is the average porosity,

S_w is the average water saturation, and

R_f is the recovery factor.

Let's assume, for example, that the areal size of the field is 5 km x 3.2 km, with the block boundary running perpendicular to the long axis of the field. We would see that the areal strip of contention, based on differences between CONUS and software calculations, is 3200 m long x 9.87 m wide, or 31,584 m^2. Next, we assume an aggregate true vertical pay-zone thickness of 60 m, not an unusual sum for GOM petroleum deposits. Then assuming average porosity = 20%, average water saturation = 20%, and recovery factor = 35%, we come up with a disputed producible hydrocarbon volume of 106,122 m^3.

Now, when we convert cubic meters to barrels, using a conversion factor of 6.289 bbl/m^3, we come up with 667,403 bbl of oil. "At $25/bbl, this has a potential of becoming a $16.7 million problem," Worsnop concludes.[21]

Thus, even high-quality computational tools will lead to errors if all parties do not agree on what is to be used.

Conclusion

Where does all this information about the Earth's shape, computational surfaces, coordinate systems, and datums lead us? For the GIS user, it allows us to make informed decisions that affect the quality of the data and the integrity of the database. With enough knowledge, we will be able to:

- Ask the proper geodetic questions about internal and external data sources
- Acquire accurately georeferenced data from primary and secondary sources
- Properly document precisely defined geospatial entities in metadata file structures or other easily assessable formats
- Define the appropriate spheroid, datum, map projection, and coordinate system for a GIS project
- Help transform disparate data sources into a unified system that can be converted from datum to datum and projection to projection
- Understand the limitations of primary and secondary source data as they relate to map scales and projections
- Qualify the level of accuracy for GIS analyses when conducting computations on geographic patterns and relationships
- Understand when it is time to bring in a professional geodesist before a major mistake occurs

References

[1] Rayson, M. 1998. "Oil Companies Underscore Importance of Geodetic Positioning." *Oil & Gas Journal* 96:27 (July 6) p. 72.

[2] Webster's dictionary, www.webster.com.

[3] Rayson, M. Oral Communication.

[4] Smith, R.S. 1997. *Introduction to Geodesy—The History and Concepts of Modern Geodesy.* New York: Wiley. p. 35.

[5] Ibid. p. 103.

[6] Rayson, M. Oral Communication.

[7] Chou, Y.H. 1997. *Exploring Spatial Analysis in Geographic Information Systems.* Santa Fe, NM: OnWord. p. 98.

[8] Morgan, J.G. 1987. "The North American Datum of 1983." *Geophysics: The Leading Edge of Exploration* (January) p. 29.

[9] Rayson, M. 2000. *Geodesy InQuest* CD ROM, Ver. 2.0.

[10] United Kingdom Offshore Operations Association. *UKOOA Guidance Notes.* www.ukooa.co.uk.

[11] Morgan, p. 29.

[12] Ibid. p. 20.

[13] Rayson, M. Oral Communication.

[14] Ibid.

[15] Rayson, M. 1998.

[16] Ibid.

[17] Burrough, P.A. and R.A. McDonnell. 1998. *Principles of Geographic Information Systems.* Oxford University Press: Oxford. p. 78.

[18] Rayson, 1998.

[19] Ibid. p. 74.

[20] Worsnop, P. Oral Communication.

[21] Ibid.

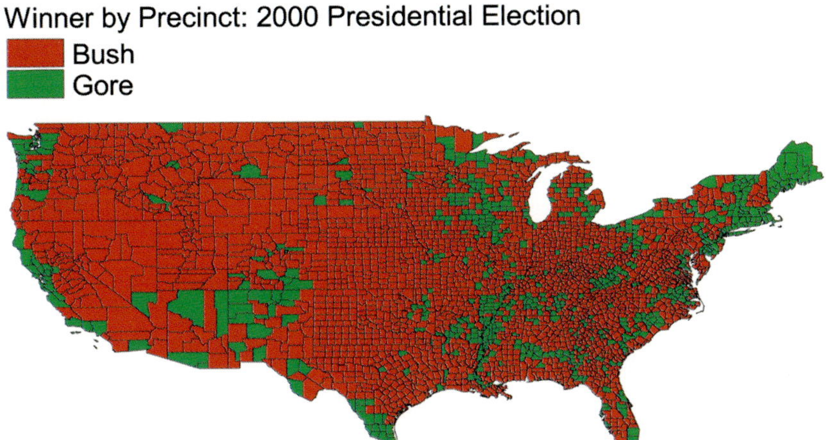

Fig. 1–6 (Plate 1–1) Spatial Comparison of 2000 Presidential Election (source: adapted from ESRI website-Election Data Services and USA Today[13])

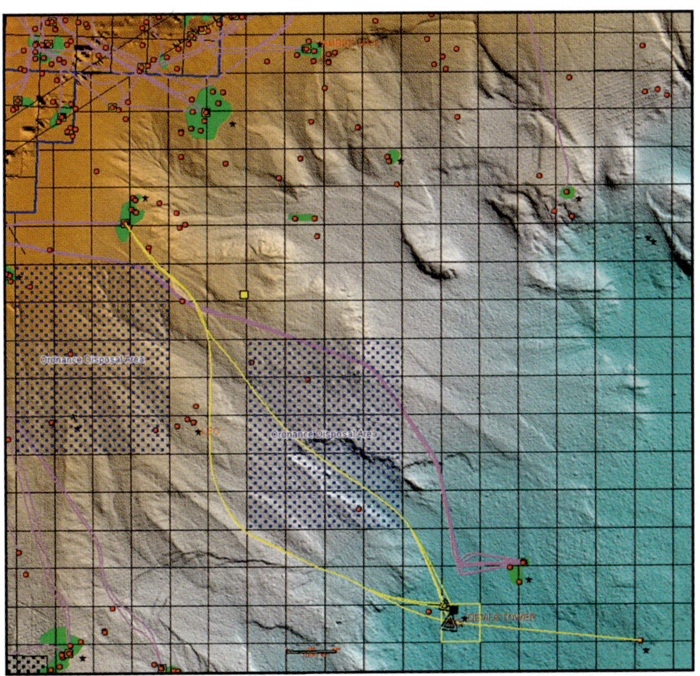

Fig. 2–3 (Plate 2–1) Devil's Tower—Planning a Pipeline Corridor (source: *Oil & Gas Journal*,[9] Dominion E&P Inc., and Geoscience Earth and Marine Services Inc.)

Fig. 2–4 (Plate 2–2) Transneft Pipeline Network
(source: data from Petroleum Argus, Wood MacKenzie, *Oil & Gas Journal*, GlavNIVC, Russian Ministry of Natural Resources, and USGS)

Fig. 2–5 (Plate 2–3) 2001 Russian Refinery Product Output
(source: data from Petroleum Argus, Wood MacKenzie, *Oil & Gas Journal*, GlavNIVC, Russian Ministry of Natural Resources, and USGS)

Fig. 2–6 (Plate 2–4) 2001 Russian Refinery Utilization and Product Price (source: data from Petroleum Argus, Wood MacKenzie, *Oil & Gas Journal*, GlavNIVC, Russian Ministry of Natural Resources, and USGS)

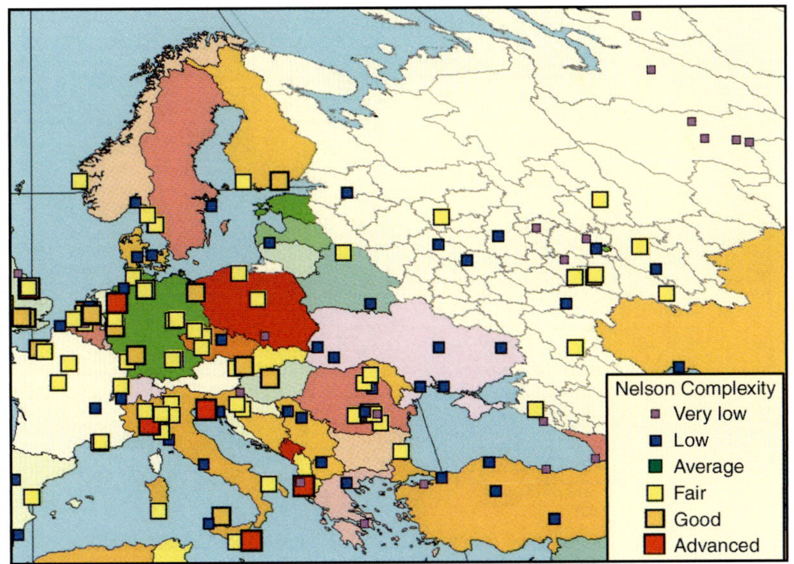

Fig. 2–7 (Plate 2–5) Nelson Cost Index for European and Russian Refineries [source: *Oil & Gas Journal* On-line Research Center (www.ogjresearch.com), Daniel Johnston Nelson Cost Index Analysis, and the *Oil & Gas Journal 2001 Annual Worldwide Refining Survey*]

Fig. 2–8 (Plate 2–6) Distance Measurement across Western Russia

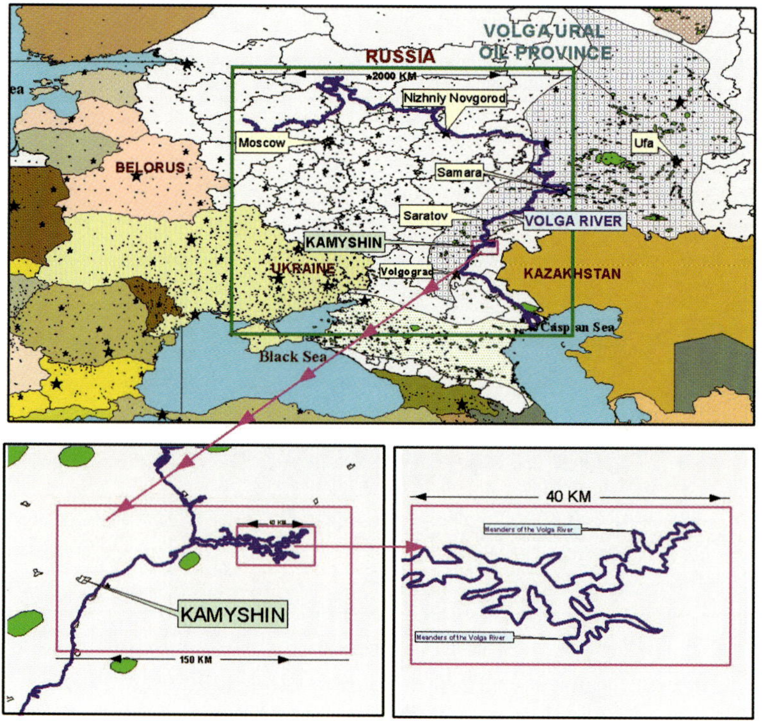

Fig. 3–1 (Plate 3–1) Effects of Map Scale

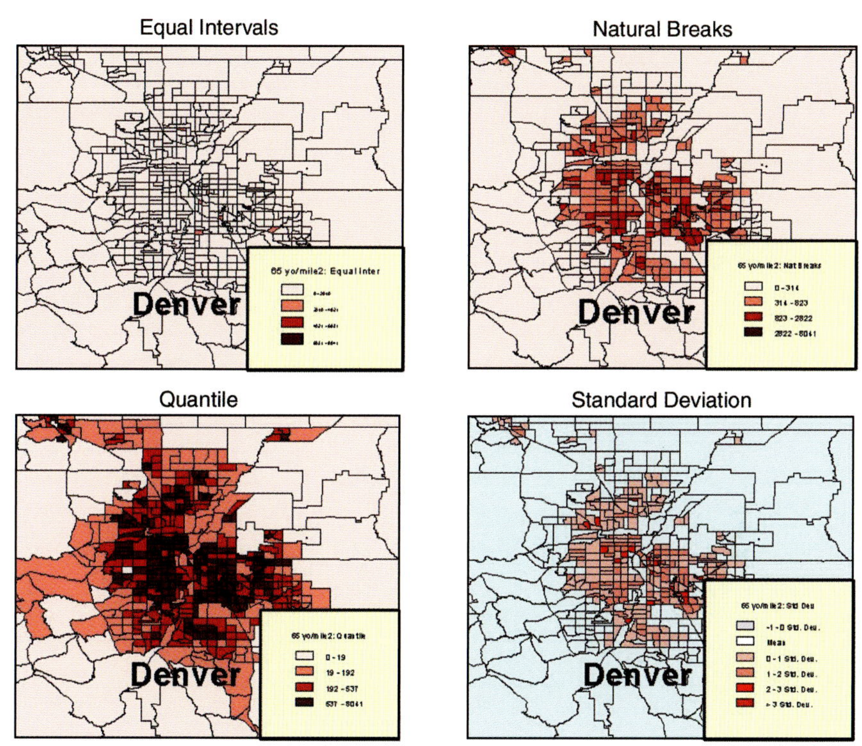

Fig. 3–5 (Plate 3–2) Four Statistical Views (data source: 2000 U.S. Census, Arcview 3.3 ESRI Data & Maps Media Kit)

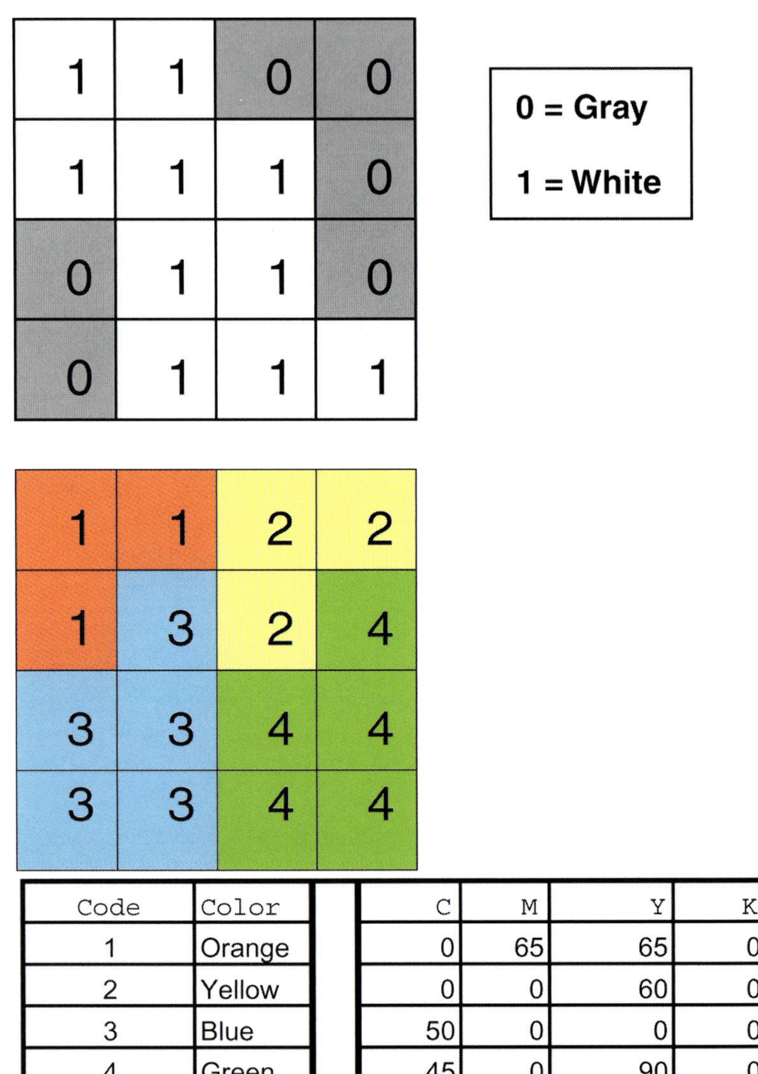

Code	Color		C	M	Y	K
1	Orange		0	65	65	0
2	Yellow		0	0	60	0
3	Blue		50	0	0	0
4	Green		45	0	90	0

Fig. 3–7 (Plate 3–3) Raster Representations

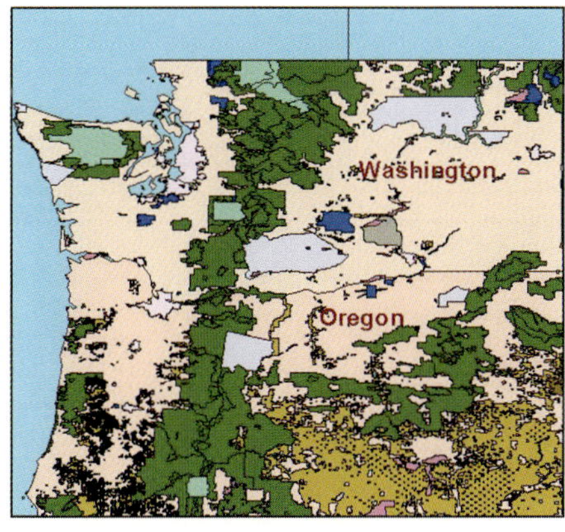

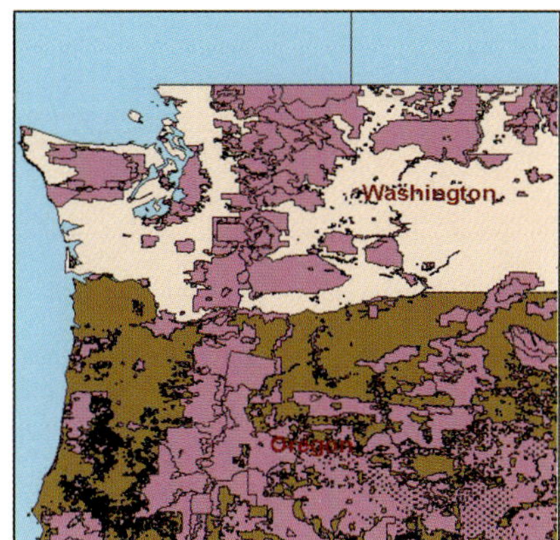

Fig. 3–11 (Plate 3–4) GIS Merged View of Washington (source: ESRI data and maps)

Fig. 4–2 (Plate 4–1) Potential for International Disputes over Resource Development [source: courtesy Hal Palmer (MaritimeBoundaries@mrj.com)]

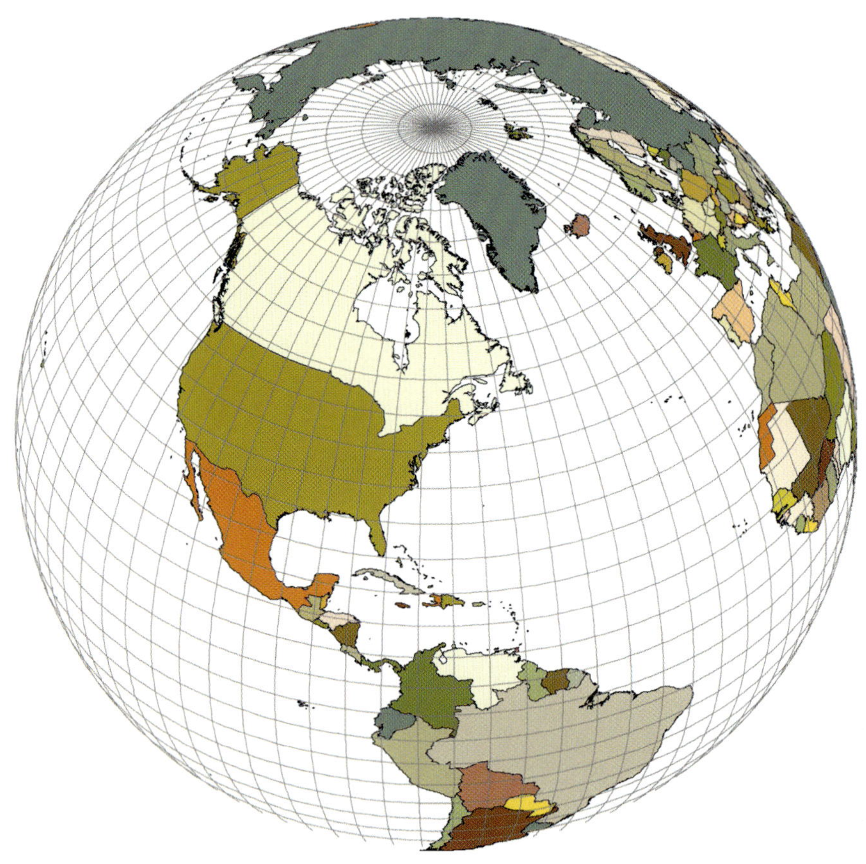

Fig. 5–4 (Plate 5–1) Two-Dimensional Depiction of the Earth from Space

Fig. 6–2 (Plate 6–1) GPS (NAVSTAR) Satellite (source: courtesy of GPS World)

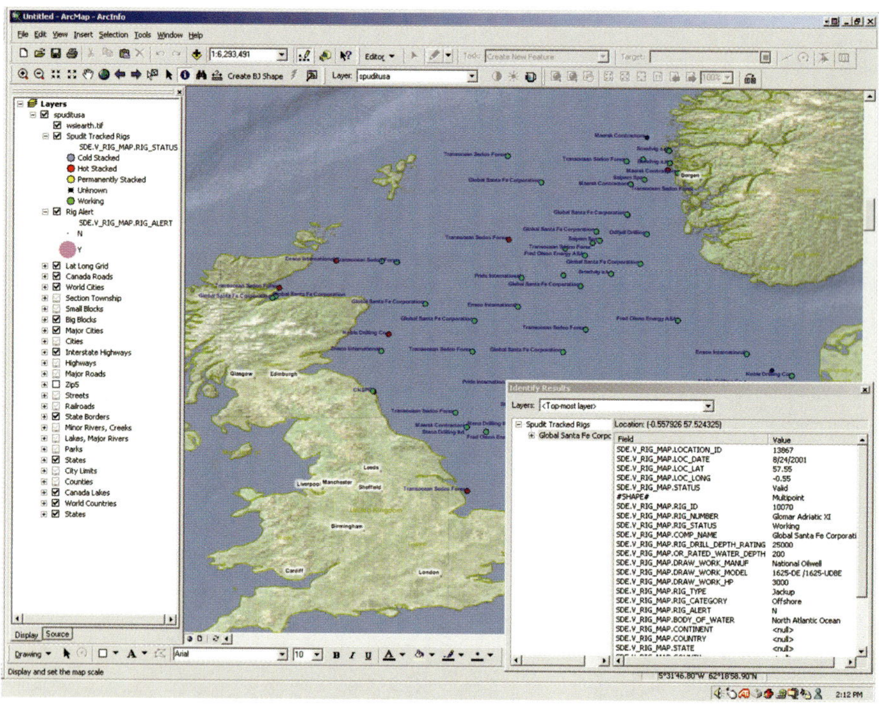

Fig. 6–4 (Plate 6–2) Drilling Rigs Locations Tracked with GPS Sensors (source: courtesy SpudIT LLC)

Fig. 7–1 (Plate 7–1) Synthetic Stereo Pair (source: courtesy Floyd Sabins[3])

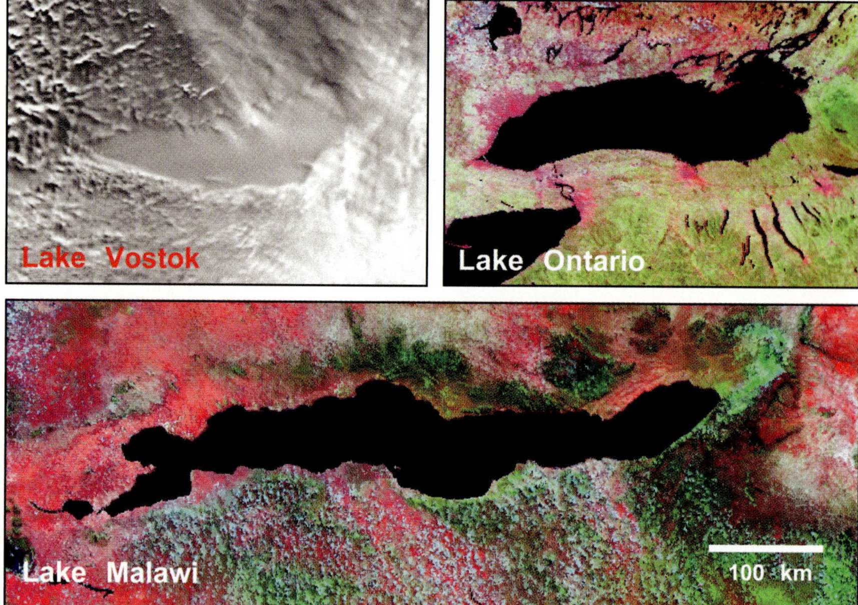

Fig. 7–3 (Plate 7–2) Determining Extent of Lake Vostok, Antarctica (source: courtesy of Dr. Michael Studinger, Lamont-Doherty Earth Observatory of Columbia University, Palisades, NY)

Fig. 7–6 (Plate 7–3) Landsat TM 2-4-7 Subscene, Saharan Atlas Mountains, Algeria (source: courtesy Floyd Sabins[21])

Fig. 7–7 (Plate 7–4) SPOT XS Infrared Color Image, Djebel Amour, Algeria (source: courtesy Floyd Sabins[24])

Fig. 7–8 (Plate 7–5) IKONOS Image of Oil Spill (source: Space Imaging Inc.)

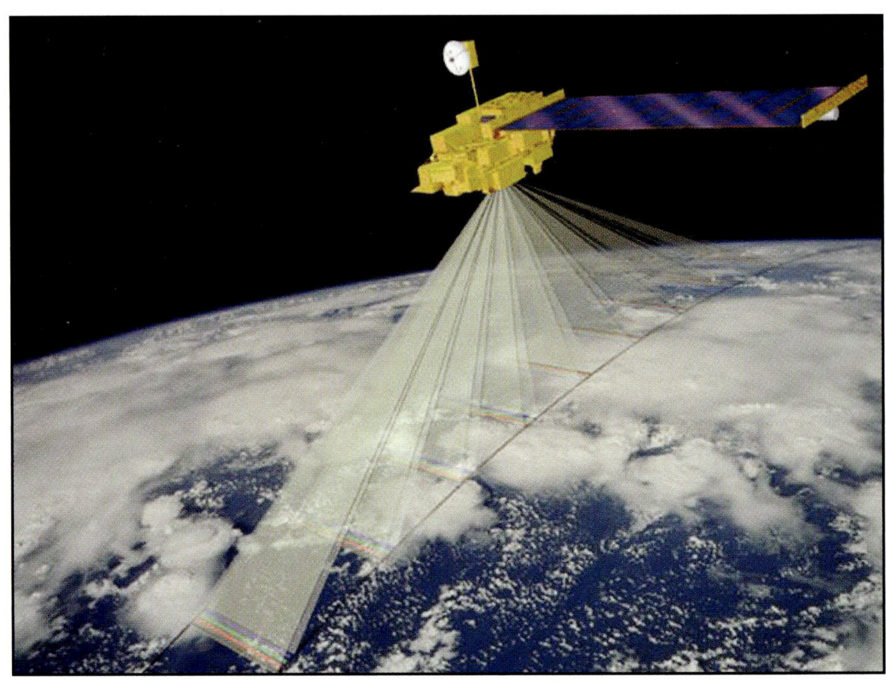

Fig. 7–9 (Plate 7–6) Earth Observing System Terra, EOS AM-1 NASA (source: F. Bordi, S. Neeck, and C. Scolese of Computer Sciences Corp., and NASA/Goddard Space Flight Center[28])

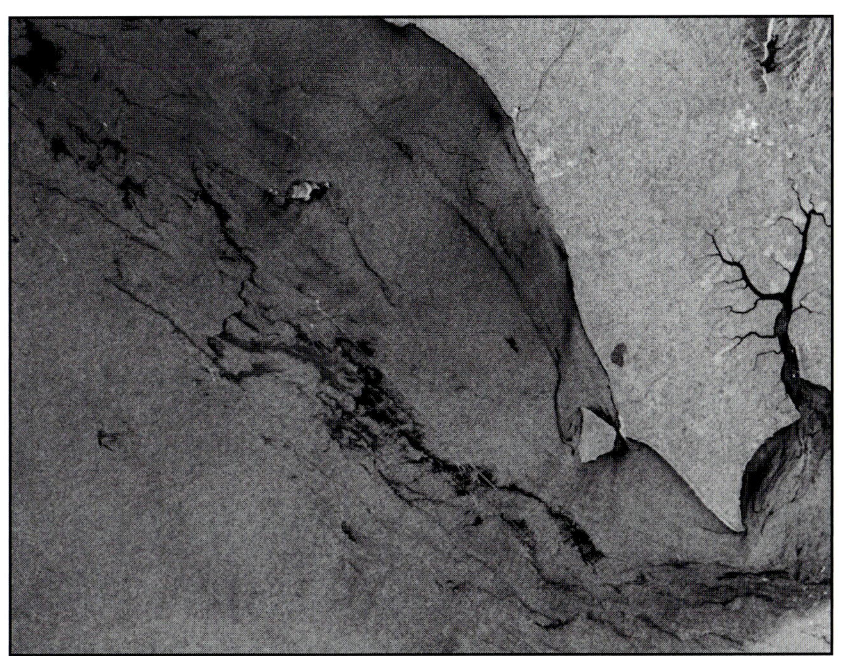

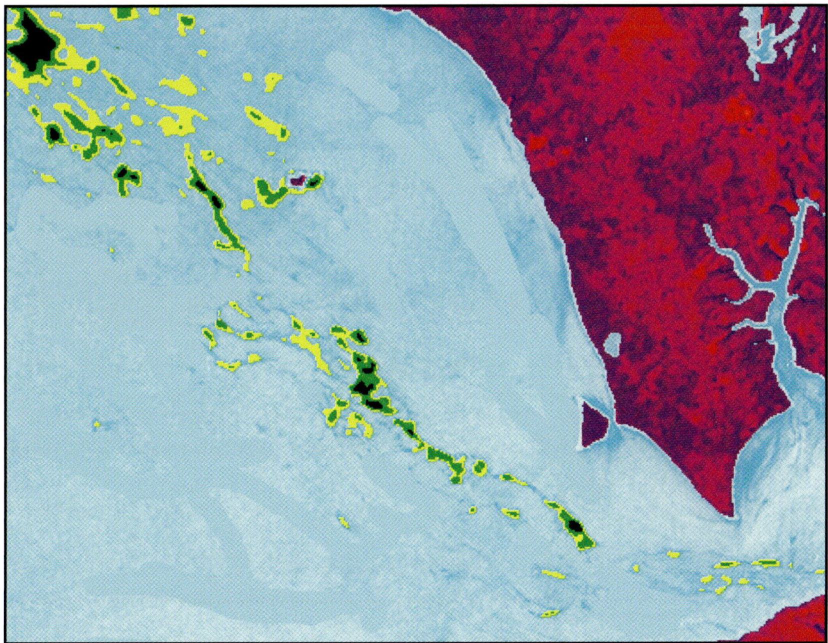

Fig. 7–11 a, b (Plate 7–7) RadarSat SAR Images
(source: Shattri Mansor, University of Putra Malaysia, Malaysia)

Fig. 8–4 (Plate 8–1) Map View Structure (sources: Petroleum Argus, Wood MacKenzie, GlavNIVC, Russian Ministry of Natural Resources, USGS)

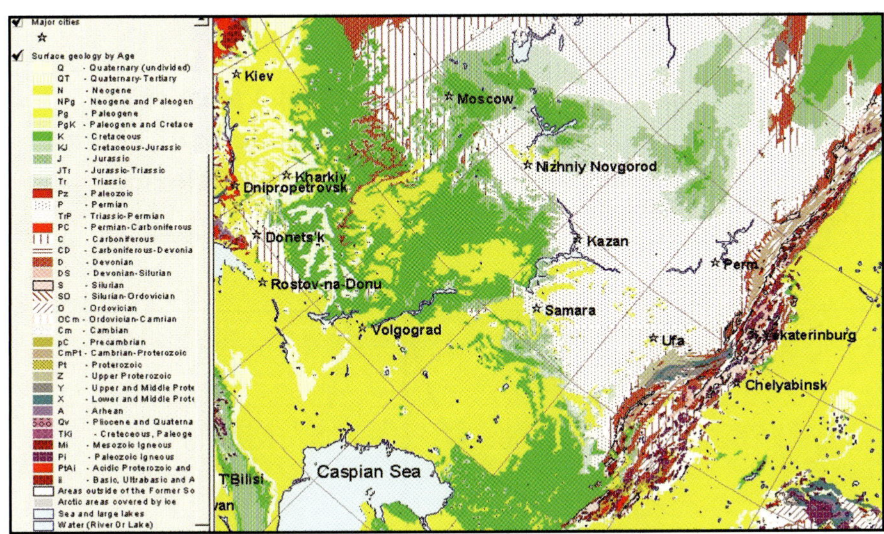

Fig. 8–6 (Plate 8–2) Distinguishing Formation Eras and Periods (source: GlavNIVC, Russian Ministry of Natural Resources, USGS)

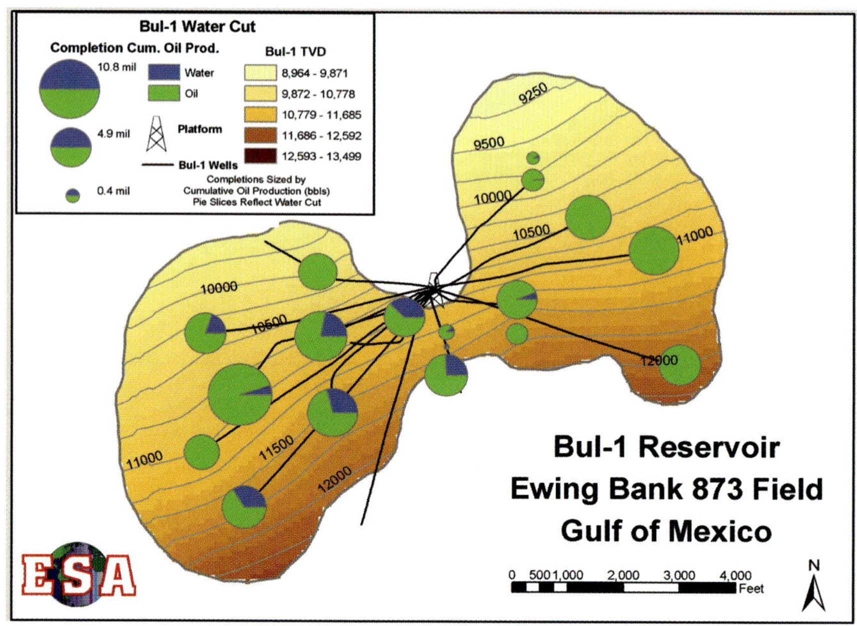

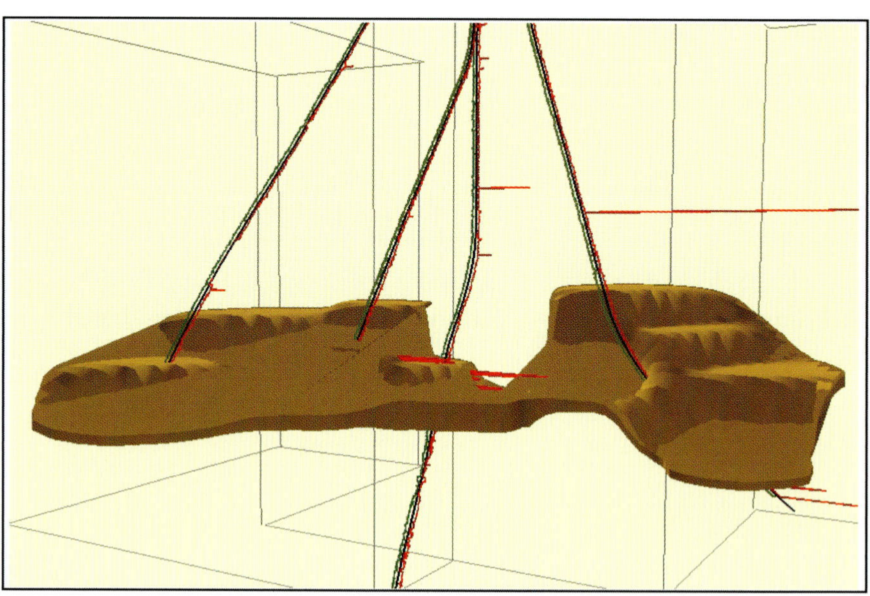

Fig. 8–8 (Plate 8–3) (a) Structural Contour Map Using GIS Symbolization Tools; and (b) Rotated 3-D View (source: courtesy John Grace, Earth Science Associates, Inc.)

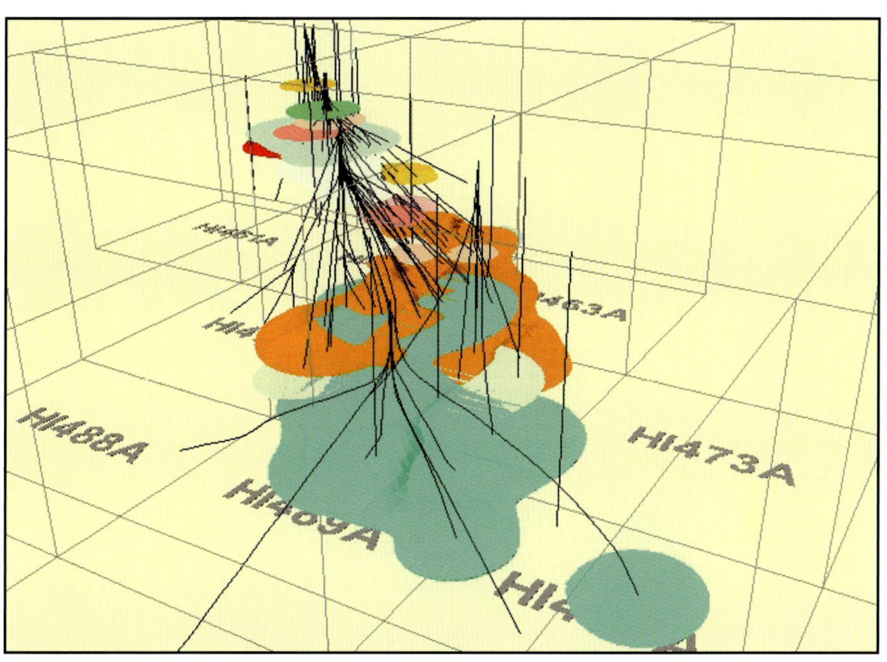

Fig. 8–9 (Plate 8–4) High Island 474A Field, 3-D View
(source: courtesy of John Grace, Earth Science Associates, Inc.)

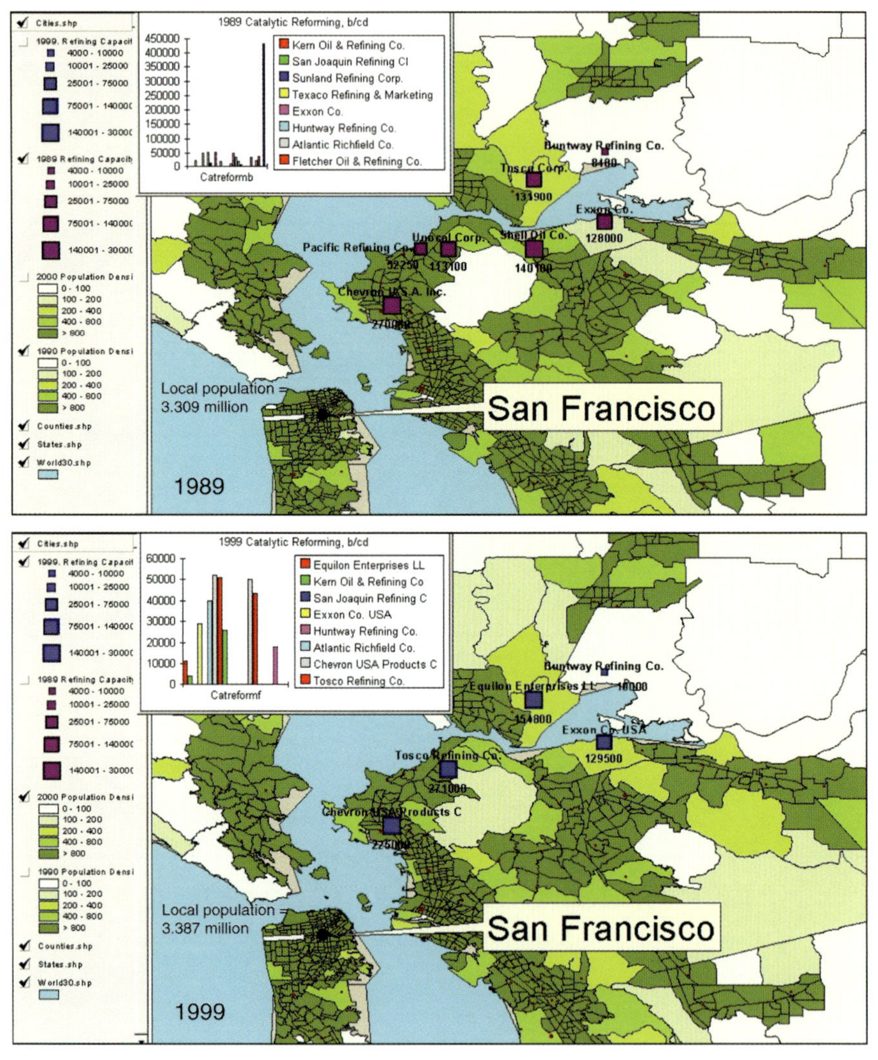

Fig. 8–10 (Plate 8–5) San Francisco Area Refineries, 1989–1999 (source: data from Oil & Gas Journal Energy Database and ESRI Data Maps and Census Tracts)

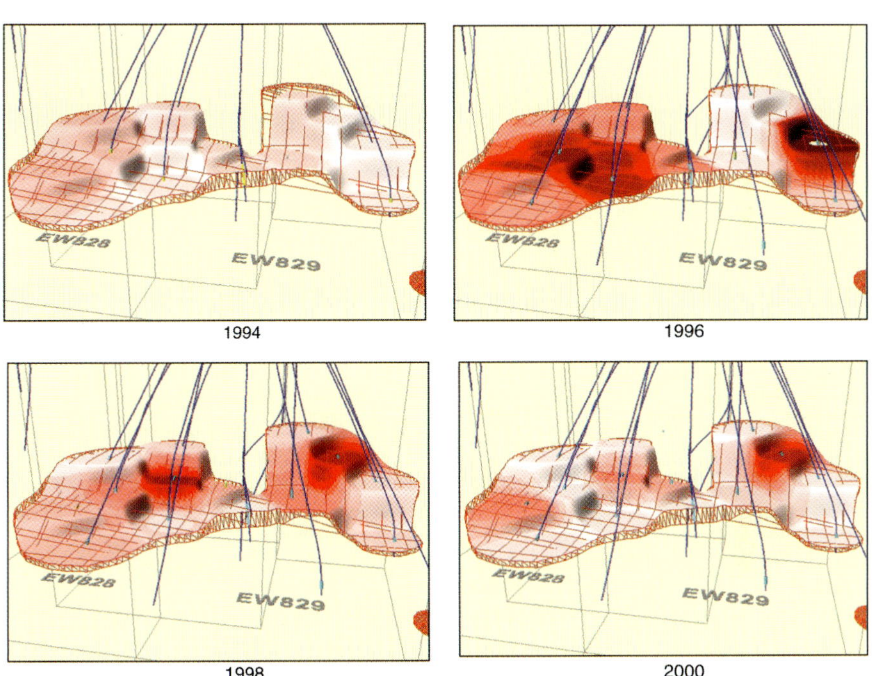

Fig. 8–11 (Plate 8–6) Time-lapse Views of Ewing Field Showing Gas Production Variations (source: courtesy John Grace, Earth Science Associates, Inc.)

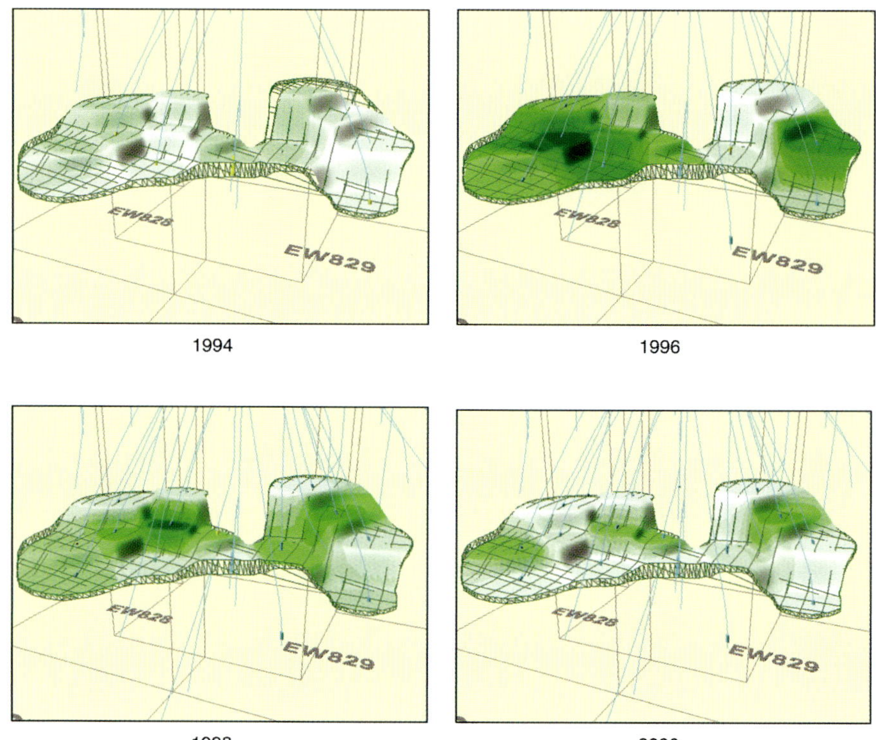

Fig. 8–12 (Plate 8–7) Time-lapse Views of Ewing Field Showing Oil Production Variations (source: courtesy John Grace, Earth Science Associates, Inc.)

5

GIS Projections

In chapter 4, we summarized the basic concepts of geodesy and how computational surfaces allow us to locate any feature on the Earth. To transfer this information onto a medium suitable for human interpretation (a 2-D surface), however, we next must apply mathematical transformations of the Earth's curved surface. This is where *map projections* and *grids* come into play.

Yet the way in which cartographers and other professionals have traditionally portrayed and analyzed geographic information may not be the best means to tackle problems with GIS. Certainly, fundamental geodetic principles continue to guide our work, perhaps more so now than ever. Yet we must learn how to separate the functions of geospatial analysis from map creation. In other words, if we wish to make a map for communication purposes, we should use a map projection. But if we wish to precisely measure area, we should use the ellipsoid. In essence, a paradigm shift is upon us, and we must understand how to best apply geodetic and cartographic techniques in unprecedented ways.

Traditional Mapmaking

Through mathematical manipulation, geodesists have developed about 200 derivative projections, based upon three developable surfaces—the cylinder, cone, and plane.[1, 2] Most of these projections only address a singular characteristic, be it area, shape, direction, or scale (distance). Because of the exclusive nature of each projection, cartographers must therefore selectively choose the most appropriate projection for the problem at hand.

For example, if you wish to show a company asset map, you would probably select a *true-shape* projection. This would show areal relationships among producing fields, pipelines, and refineries. Or if you need to accurately measure pipeline length, you would probably choose a *true-scale* map. You may even be asked to determine the ideal spot for a gasoline station. You could then build an *equal-area* real-estate map with population density shown as a function of income. The problems are endless. However, the basic concepts can be applied to each and every project. Let us begin by looking at those factors that limit and augment the art of traditional mapmaking.

Distortion

A map projection is a systematic representation of all or part of the surface of a round body on a plane. Unfortunately, any time we try to represent a curved surface on a 2-D surface, we introduce complications. These include instances of rotation (change in orientation), translation (change in position), and distortion (change in shape). (For most of the book, the term *distortion* will be used to represent all three instances.) The easiest way to visualize this is to peel an orange and press it flat on a table. For a very small piece, changes in shape, orientation, and position will be small. For a large piece, however, you will notice it is impossible to press it flat without breaking the skin.

Thus, we must accept the fact that there is no way to smooth out a large spherical surface and keep intact all the useful features we wish to portray.[3] It then becomes a decision of which characteristic we wish to maintain. For example, we may need to keep a correct perspective for shape, but in doing so, must sacrifice scale (distance). Or we may wish to maintain area at the expense of shape. Or in other cases, we may wish to produce compromises among all characteristics.

We can keep our scale of area constant all over the map—at a price. But we can't keep the scale of distance constant all over the map at any price.

— David Greenwood

Projection Surfaces

As mentioned, the three basic projection surfaces are the cylinder, cone, and plane. To imagine how features from a sphere, such as a globe, can be projected onto a cylinder, let's conduct the following classic experiment:

1. Set up a transparent globe with only latitudes and longitudes (every 30°) outlined.
2. Install a center-fixed light within the globe so that it shines outward in all directions.
3. Stand a piece of paper upright and wrap it around the globe so that it contacts the equator completely around its circumference (see Fig. 5–1).

Now with the light turned on, notice how the network of geographic lines is projected onto the piece of paper as shadows. Then imagine that the paper cylinder can preserve all these lines as a negative imprint. As the paper is cut, unwrapped, and laid flat on the table, the spacing of the parallels (latitudes) and meridians (longitudes) will be clearly seen.

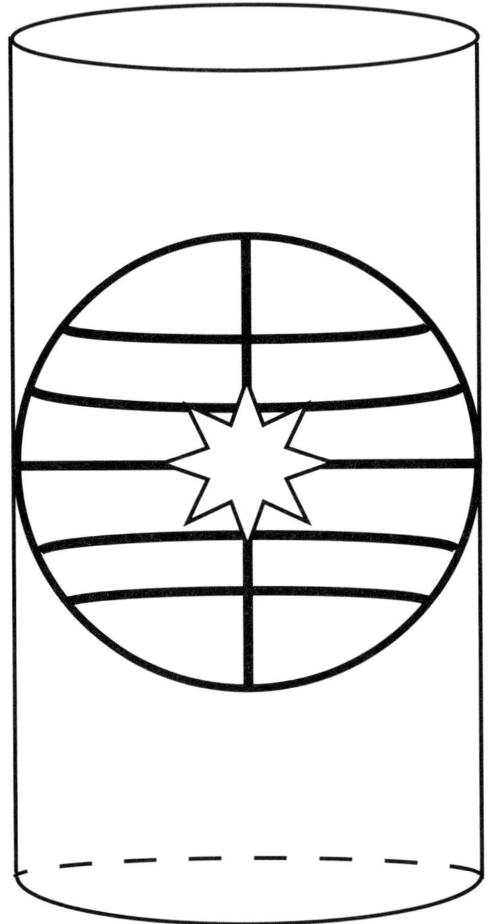

Fig. 5–1 Classic Upright Cylinder with Center-Fixed Light

At this point, notice how the parallels of latitude, which on the globe form equidistant lines, are marked on the paper with ever-increasing separation from the equator. This basic projection is commonly called a *simple perspective cylindrical projection* (see Fig. 5–2).

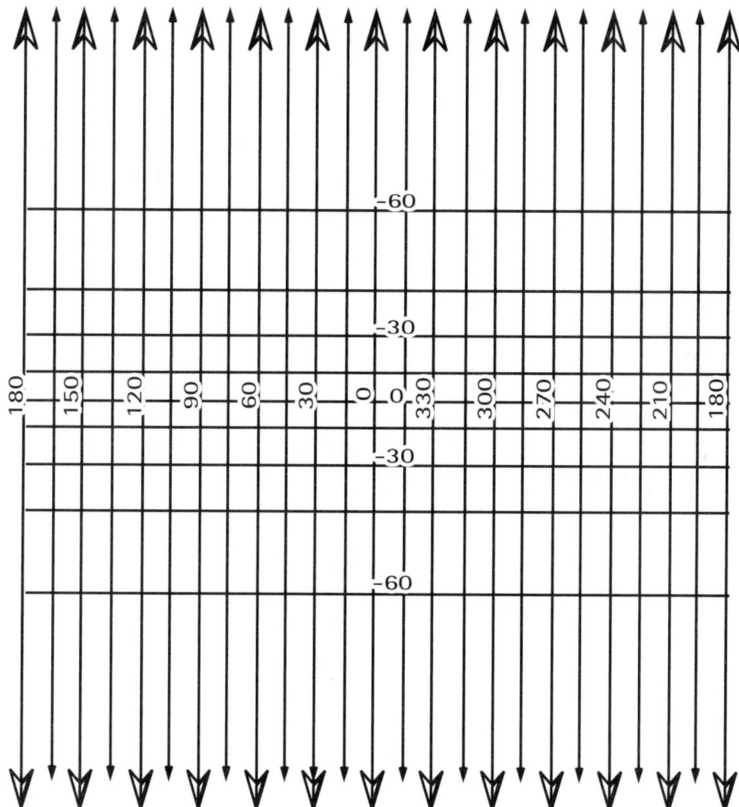

Fig. 5–2 Simple Perspective Cylindrical Projection

A more famous cylindrical projection is the *Mercator*, developed more than 400 years ago by Gerhardus Mercator (see Fig. 5–3). This map served one specific purpose. It allowed sailors to chart a straight-line directional course (rhumb lines) to their destination instead of plotting a complicated series of ever-changing courses along a curved path. Mercator built this map by stretching latitudes north and south of the equator by a ratio equal to the equator's circumference divided by the length of each latitude.[4]

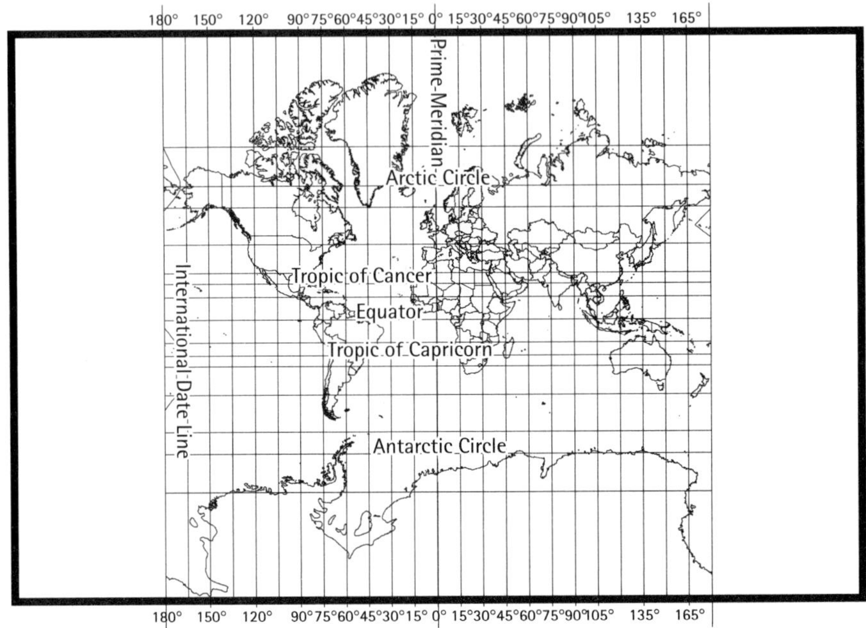

Fig. 5–3 The Mercator Projection

To visualize this, compare what the 60th parallel looks like on both a Mercator map and a globe [see Fig. 5–3 and Fig. 5–4 (Plate 5–1)]. Notice that its circumference is one-half that of the globe's equator, but of equal length on the Mercator projection. To achieve conformality, or the maintenance of shape, Mercator had to double the circumference of the 60th parallel on his map. The result is a geographic system of parallel longitudes.

To visualize this, again take the globe and compare what Greenland looks like on the Mercator map. It will readily be noticed that the basic shape is correct on the Mercator projection. However, notice that areas become increasingly exaggerated as distance increases from the equator. For instance, Greenland is shown larger than South America, although in reality, it is only one-eighth the size. (The Mercator does preserve features exceptionally well along the equator.)

Conic projections

Next, let us examine what takes place when we project the latitudes and longitudes onto a cone. Whereas cylindrical maps provide excellent views of the world, conic projections provide great utility for portraying continental, regional, and local areas. This is true especially for features with elongated east-west extents. Conic projections, on the other hand, become quite useful for countries found in middle latitudes, such as North America, Europe, and Russia.

To see what happens when latitudes and longitudes are projected onto a cone, we again take the globe and light setup. But this time, a conical "hat" is placed over the globe, with its peak positioned along the polar axis of the Earth. The surface of the cone should be touching the globe along some particular latitude (known as the *standard parallel*), much like a sweatband. Then we can again trace the shadows of the latitudes and longitudes onto the paper.

Once we slit the cone from apex to rim and lay it flat, we find that we have made a fan. The meridians are projected onto the cone as equidistant straight lines radiating from the apex (see Fig. 5–5). Additionally, the parallels will be marked as lines around the circumference of the cone in planes perpendicular to the Earth's axis.[5]

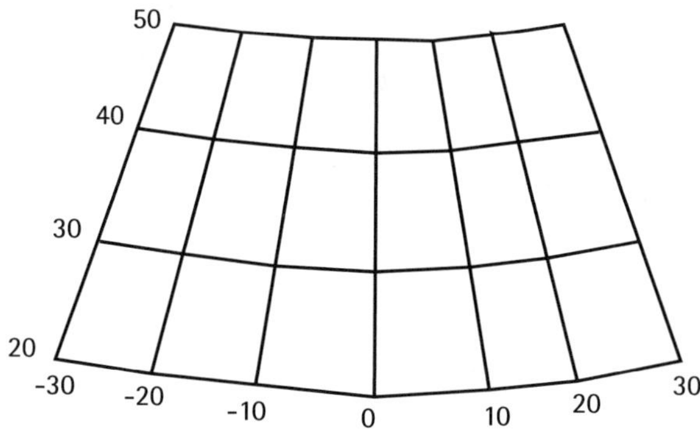

Fig. 5–5 Conic Fan Shape

Conic projections can be modified to produce equidistant, conformal, and equal-area maps by the strategic arrangement of one or two standard parallels. To do this, place two secant lines along two meridians (parallels), so as to undercut the globe in a precise way (see Fig. 5–6). Now we can carve out cross sections over the area of choice and produce conformal or equal area projections.[6] It should be noted that plotting surfaces may be tangent or secant to the model of the Earth. Tangency refers to a plane that touches a surface at a single point. Secancy, however, takes a slice out of a surface along two points. (If tangent lines were used, there would only be one standard parallel.)

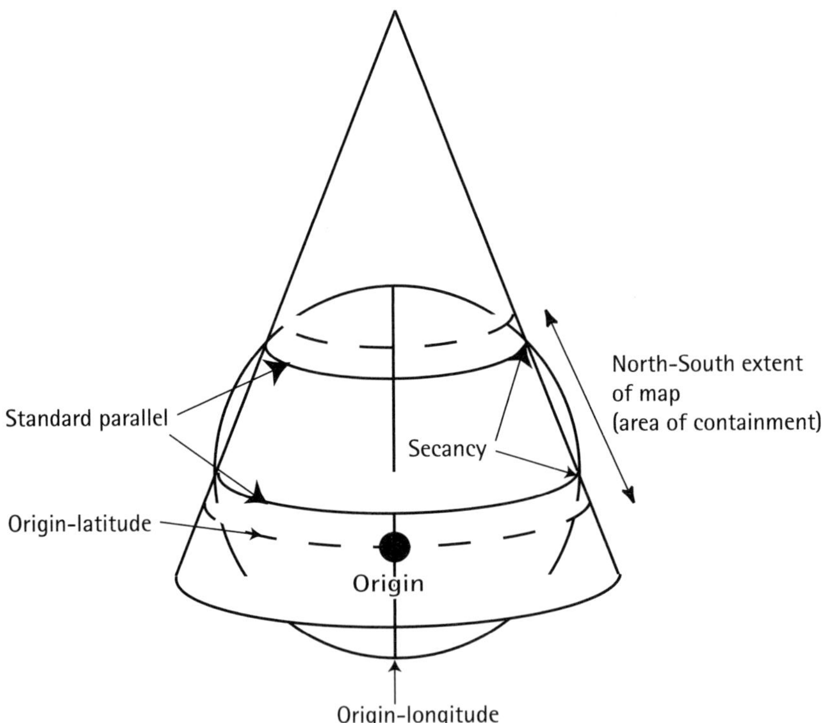

Fig. 5–6 Conical Development of Lambert Conformal and Albers Equal-Area Projections

Johann Heinrich Lambert and Heinrich Christian Albers applied this technique in the 18th and 19th centuries, respectively, producing two widely used projections. These were the *Lambert conformal conic* and *Albers equal-area conic* projections.[7] For these projections, areas that lie outside the *limits of the projection*, or outside the bounding parallels, will realize distortion. Features within will maintain reasonably good shape and areal integrity.[8]

The Albers equal-area conic projection has especially found great use in regional maps of the United States. This is because the properties of area, shape, and distance remain within reasonable limits. (The Albers equal-area conic projection, with standard parallels positioned at 29½° and 45½°, is the system of choice for most U.S. maps of the *National Geographic*.)

And where the Mercator projection makes a scale error of 40% at the northern border of the United States, Lambert's conformal conic projection, with parallels situated at 33° and 45°, is off scale only 0.51% from the Atlantic Coast to the Pacific Coast.[9] Thus, it is easy to see why the USGS uses the Lambert conformal conic projection for its 7½- and 15-minute quadrangle maps for 32 U.S. states. If one measured the straight, true distance between New York and San Francisco, for example, its scale would be short by only 12 mi.

Grids

Now that we have some base knowledge of how ellipsoids, datums, and projections work, we next turn to the subject of base-map grids. These are used by the American, European, Russian, and Chinese petroleum industries. A grid consists of intersecting lines that represent northing and easting departures from some point of origin (see Fig. 5–7). It is the most convenient way of representing positions on a map.

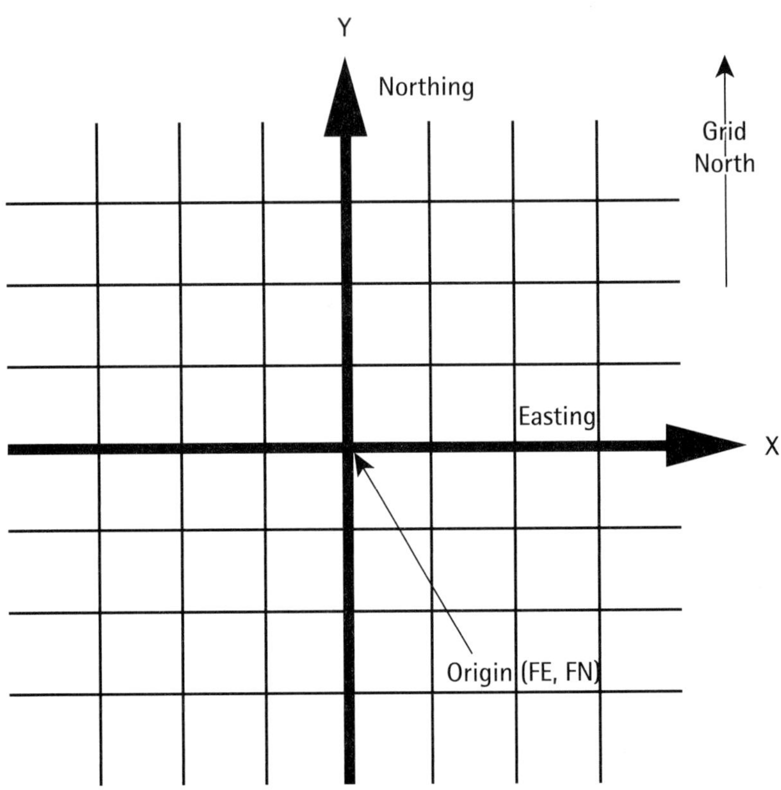

FE = False Easting
FN = False Northing

Fig. 5–7 Typical Grid System

The base-map grid allows cartographers a way to translate 3-D latitudes and longitudes onto a system that allows direct 2-D measurements in meters, feet, links, chains, or other units. Thus, we can precisely locate points in direct relation to one another.

The downside to a grid system, as compared to geographic coordinates, however, is there is no way to determine grid locations independently unless special attention is paid to zonal distinction. In other words, we can have two numerically identical X and Y coordinates that can still be spotted on two different base maps. For example, 20 ft north and 30 ft east of Farmer Jones's most southwest property corner is not the same thing as 20 ft north and 30 ft east of Farmer Smith's. This can lead to confusion unless the grid zone is classified.

Regardless, grid systems have the following advantages:

- Cartesian coordinates may be displayed on a flat surface.
- Measurements of area and distance can be easily calculated on a 2-D surface.
- The user can apply simple planar trigonometric and geometric calculations directly on the map, mathematically, or within the GIS.

Warsaw and NATO grids

Two of the most widely used grid systems include the *Warsaw Pact Gauss Kruger* (GK) and NATO's *universal transverse Mercator* (UTM). A comparison between the two shows the importance of Cartesian coordinate frameworks in cartographic work.

To understand how a transverse cylindrical projection is conceptualized, imagine how the simple perspective cylindrical projection was derived using the light and globe setup. This time, however, orient the paper cylinder 90° on its side so that the top of each pole tangentially touches the cylinder in a horizontal perspective. To produce the universal transverse Mercator (as used by NATO), we take this one step further and decrease the size of the cylinder through secancy (see Fig. 5–8). In this way, the cylinder slices through the globe with both poles falling outside the cylinder. Note that the UTM cylinder undercuts the globe through secancy, thereby limiting the Earth to latitudes between 84° N and 80° S. Conversely, the GK remains tangent at both poles.

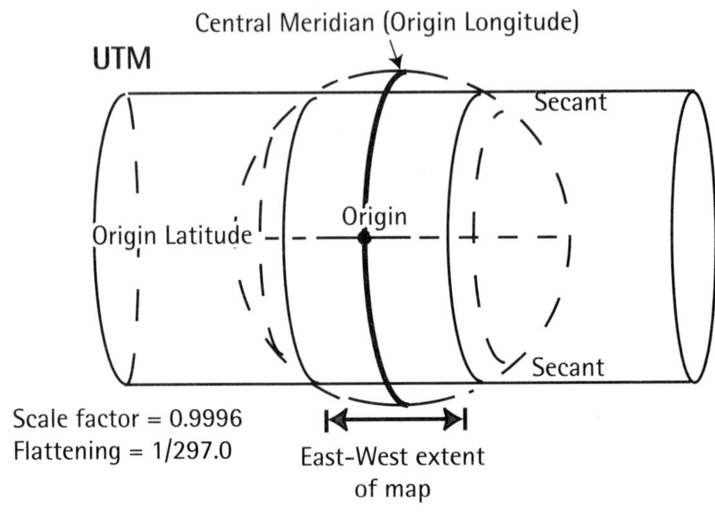

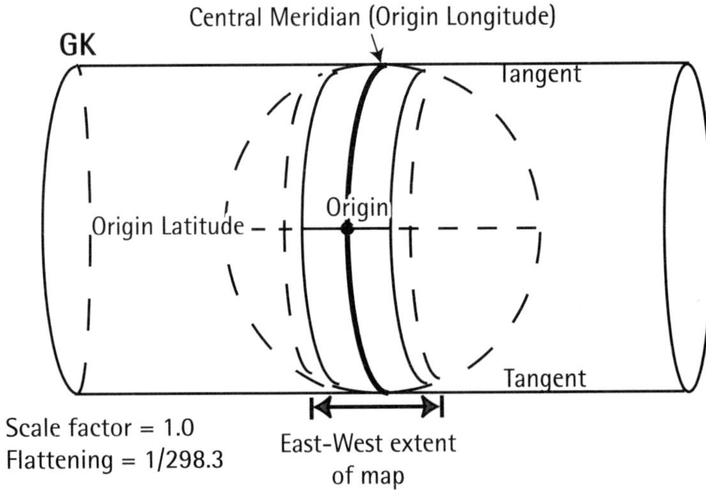

Fig. 5–8 Differences between UTM and GK Grid Systems

The GK and UTM grids have similarities and differences. These are functions of the ellipsoid, datum, scale factor, central meridian, false northings and eastings, and north-south measurement systems.

First, the GK uses the Krassovsky ellipsoid, while the European version of the UTM uses the Hayford. Thus, both ellipsoids use different models for the shape of the Earth. (The ellipsoidal parameters published by Feodosi N. Krassovsky in 1942 have proven to be remarkably close to satellite geodetic methods.)

Second, the GK datum is tied to the Pulkova Observatory near St. Petersburg, Russia, while the European UTM uses Potsdam, East Germany. Thus, each system uses a different reference point for ground control measurements.[10]

Third, both systems divide the globe into 60 narrow longitude zones measured in increments of 6°. With the GK grid, however, Zone 1 begins at Greenwich, whereas the UTM grid begins at 180°. In Figure 5–9 the 6° grid zones are shown from a flattened perspective. (The Chinese also use the GK projection. As a result, the use of GK equals or perhaps surpasses the use of UTM.)

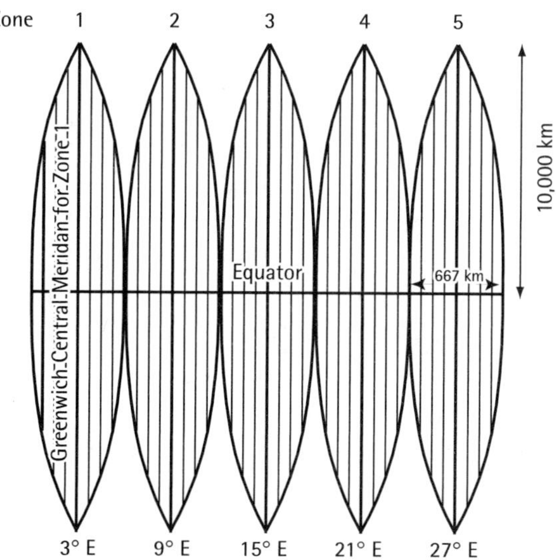

Fig. 5–9 Flattened Perspective of GK Grid (source: modified from U.S. Dept. of the Army Field Manual)[11]

Both grids, however, use a central meridian line that divides each 6° zone in half. "The beauty of using longitude zones 6° wide is that, at the equator, a zone is about 667 km wide. Thus, it is possible to create numbering systems so that distances on the grid in km have at most three digits."[12] (The Earth's circumference is 40,075 km divided by 60 zones.) To maintain positive values across each zone, false eastings of 500,000 m are applied to both grids. Thus, the central meridian line has a designated value of 500,000 m (see Fig. 5–10). The zones are further broken down into 100,000-m east-west subdivisions, with measurements taken from the left-hand corner of each subdivision.

Finally, with the GK grid, all zones extend from the North Pole to the South Pole in a concentric manner. (Recall that the cylinder is placed tangential to the globe on a transverse Mercator.) In comparison, the UTM grid is bounded between the latitudes of 84° N and 80° S through secancy. Thus the zone lines come to an abrupt end just short of each pole.

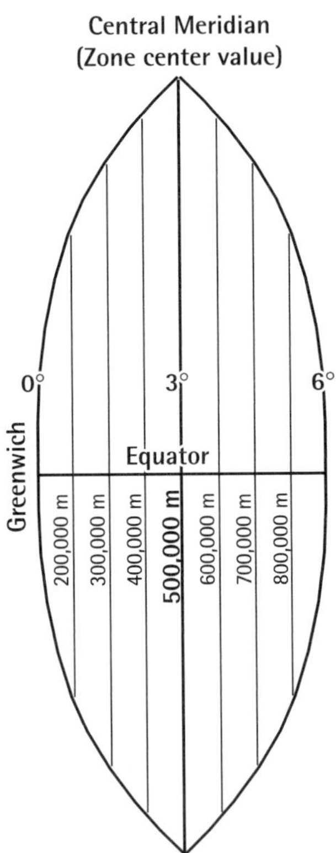

Fig. 5–10 GK Grid Zone 1 (source: modified from U.S. Dept. of the Army)[13]

For both grids, north-south measurements are given in meters. Fortunately, the metric system was originally defined so that the distance from the equator to each pole is about 10 million m (10,000 km). Although modern geodetic computations show that this isn't quite perfect, it is close enough to make maps with a high degree of accuracy. Thus, both the NATO and Warsaw Pact grids begin counting from 0 at the equator.

There are differences, though, in the way in which north-south measurements are denoted. The UTM, for example, divides the Earth horizontally into 20 zones measured in 8° increments. It begins at 80° S with the letter C and continues northward to 84° N. The letters *I* and *O* are not used, and row X is 12° high (see Fig. 5–11). False northings of 0 are assigned to UTM zones in the Northern Hemisphere. A northing of 10,000,000 m is applied to the Southern Hemisphere to maintain positive values.

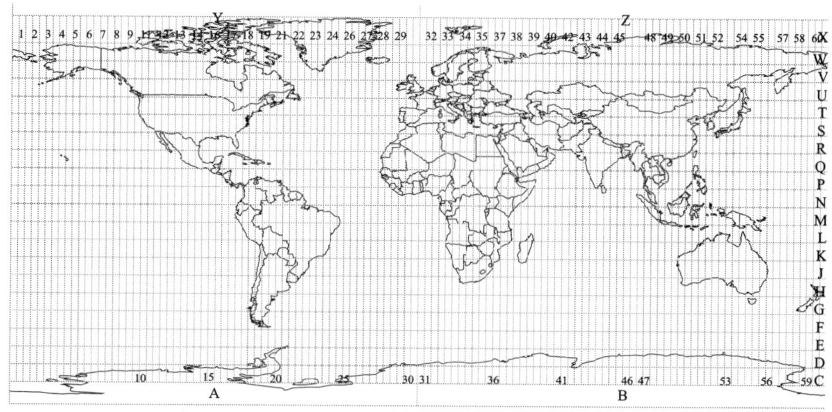

Fig. 5–11 UTM Grid Zone Designations

Conversely, the GK grid simply divides the Earth in half. A value of 0 is assigned to the equator.[14] South of the equator, a false northing of 10,000,000 m is also applied to maintain positive values.

Because the GK and UTM coordinate systems can use similar values from zone to zone, the zone number or letter designation must always accompany the coordinates. Yet again, both systems differ in the manner of description. For example, the GK grid coordinate 5850760, 3334000 is the same as UTM grid coordinate 33UU 34061, U 48440 (see Box 5–1).[15]

Box 5–1

*Coordinate Comparison of Warsaw Pact and NATO Grids**

Warsaw Pact Gauss-Krassovsky

Grid coordinate: 5850760, 3334000

y component 5850760

Northern measurement recorded directly in m from equator.

x component 3334000

First "3" stands for easting zone number, measured from Greenwich in increments of 6° (zone falls between at 12° and 18° E longitude). Second "3" stands for 100,000-m block designation."34000" stands for direct east measurement in m.

NATO UTM

Grid coordinate: 33 UU 34061, U 48440

x component 33 UU 34061

"33" stands for vertical zone number measured from 180° in increments of 6° (zone falls between at 12° and 18° E longitude). "UU" stand for easting 100,000-m block designation."34061" stands for direct east measurement in m.

y component U 48440

"U" stands for horizontal zone number (48° N latitude). "48440" stand for direct north measurement in m from bottom of zone.

*It should be noted that Russians and the Chinese read their maps up-and-right as compared to right-and-up for Westerners. This is because the GK northing element forms the first half of the grid coordinate, and vice-versa for the UTM.

Note: Adapted after Department of the Army Field Manual No. 34-85.

To unambiguously identify a UTM or GK grid, the following data must be supplied:

- Latitude of the natural origin
- Longitude of the natural origin
- Scale factor at the natural origin
- False easting
- False northing

When working with maps, no matter what vintage or origin, Rayson points out that one should never forget the datum when it comes to a grid system. "All coordinates expressed on a grid are projected from geographical coordinates, and these coordinates remain tied to the Earth through the *datum origin*." In other words, just because coordinates have been converted to a plane grid, it does not mean that they have been disassociated from the datum. "Always know the (specific) datum, because without (this benchmark), grid coordinates cannot be confidently converted back to geographical coordinates."[16]

As can be seen, understanding grid systems is very important in cartographic work. A GIS user who becomes familiar with both GK and UTM systems will have an advantage in international work. These grids are the dominant large- and medium-scale mapping grids used in the United States, Russia, Europe, and China (see Box 5–2).

Box 5–2
Map Scales

The subject of map scales can be confusing to those new to cartography. When someone refers to a map of large scale, for example, this person is simply describing features mapped within a small area. On the other hand, small scale refers to large areas. Two methods are used to describe maps scales:

1. True scale (i.e., 1:50,000).
2. Representative scale (i.e., 1 in. = 5000 ft).

True scale is unambiguous and works for all measurement units of length including Metric and English. For example, a map measured in centimeters and produced on a 1:50,000 scale simply means that for every centimeter on the map we can go out and physically measure 50,000 cm on the ground (taking care to apply the map scale factor).

An easy way to look at map scale is to interpret the numbers as a ratio and see what happens to each mathematically. For example, 1:25,000 can be interpreted as 1/25,000, which equals 0.0004. On the other hand, 1:1,000,000 can be interpreted as 1/1,000,000, which equals 0.000001. Obviously the latter is much smaller than the former.

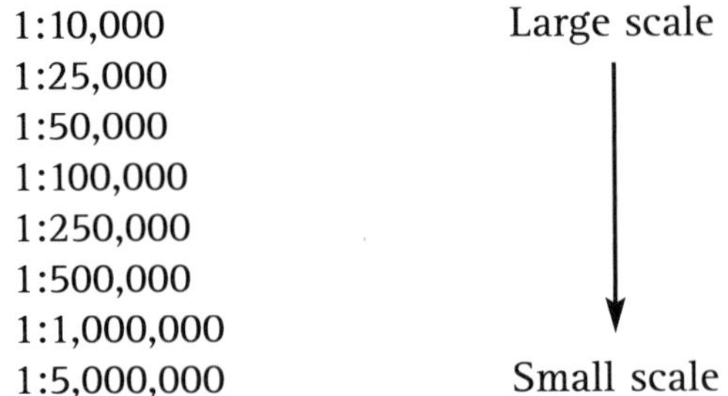

A New Way of Thinking

To portray meaningful relationships for a complex, 3-dimensional world on a flat sheet of paper or a video screen, a map must distort reality.

—Mark Monmonier

As GIS users, we must be extremely wary of the effects of distortion. One must not be lulled into thinking a software program can perform spatial calculations without breaking the laws of geodesy. As Mike McCabe, an oil industry geodesist with Enron Global E&P explains, "If data have been projected, any type of spatial work on that data will be subject to distortion inherent in that particular projection."[17]

Thus, the GIS user should not fall prey to the belief that GIS measurements are always right on the money. As he explains, "You must take care of earth-curvature....It is no good running out a 100-km pipeline, only to find that you are 33 m off trying to connect it to the platform. In such cases, the use of GIS should be used for preplanning purposes only, but never for precise geodetic positioning."[18]

Some GIS purists may dispute this, claiming that a system can be built with extreme accuracy for purposes of project implementation. However, when it comes time to connect a major pipeline system that encompasses hundreds of miles over rough terrain, there can be no substitute for physical surveys. And it is almost certain that no operator will try to complete a project without one.

Yet through the processing capabilities of the computer, GIS technologies can eliminate many shortfalls associated with cartographic measurements once conducted on hard-copy maps. For example, to measure the distance between two cities requires a properly aligned equidistant map projection, a known scale, and ruler. Yet try to use this same method to measure the length of a meandering stream. Or try to measure the exact distance between Denver and Grand Junction, Colorado, through the mountainous terrain of the U.S. Rocky Mountains.

Fortunately, when it comes to measuring distance in a GIS, we have some very powerful tools on hand. Most GIS software programs, for example, allow the user to quickly project and deproject, measure curvilinear distance, area, and even volumes. Yet because we have these tools, we must change our philosophy on how to work with geospatial data. In other words, we must apply the full potential of geodetic principles in our work.

For instance, assume we encounter huge differences of measurement from projection to projection as shown in Table 5–1 and Figure 5–12. Note the distortion in size and shape in Figure 5–12 (not to scale). Should we continue to work with a projected database that will continuously haunt us with problems of distortion? Or should we move towards systems that contain nonprojected geographic coordinates? Remember, even with an equal-distant conic projection or UTM grid, accurate distance can only be maintained along certain lines. Even then, we will only get distance right, but never area. This is especially true once measurements begin to stray from the standard parallel or point of origin.

Geographic Unprojected

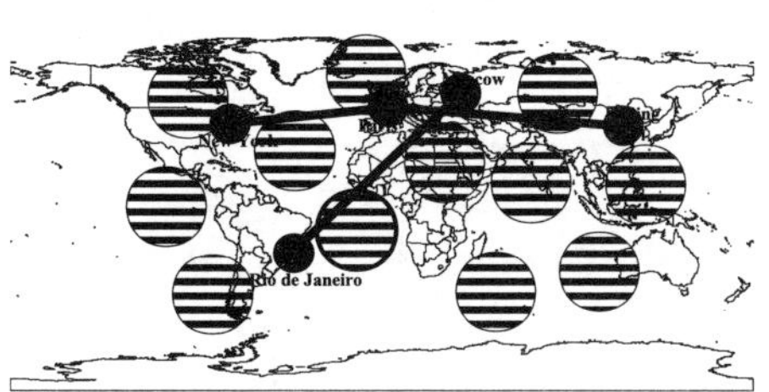

Mercator Cylindrical

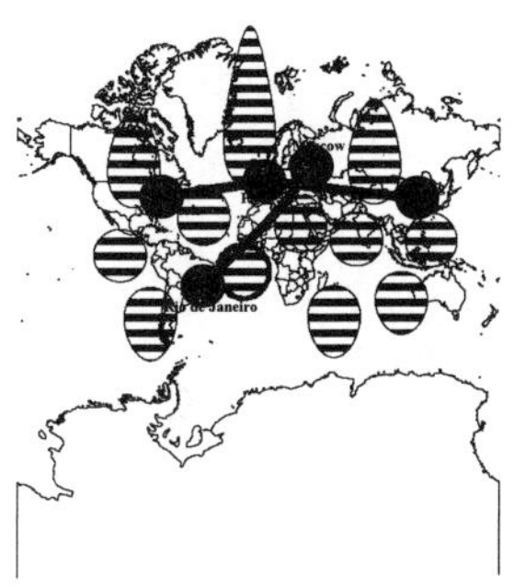

Mollweide Pseudocylindrical

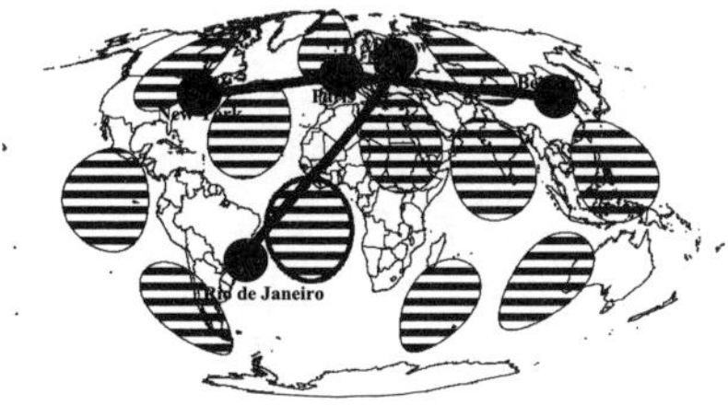

Peters Equal-Area Cylindrical

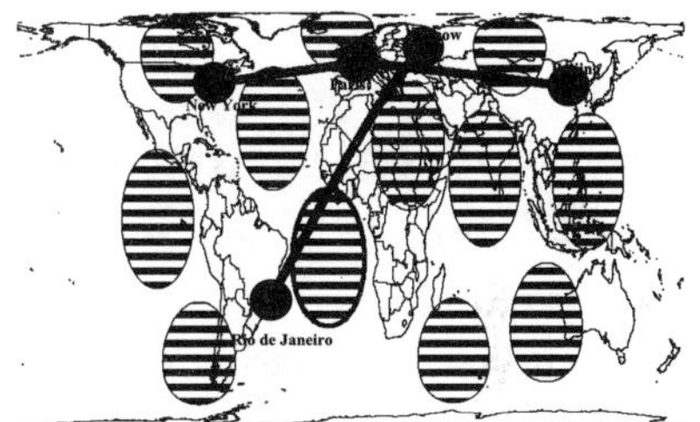

Fig. 5–12 Projection Comparisons

	New York to Paris	Paris to Beijing	Rio de Janeiro to Moscow
Geographic Unprojected	3635	5104	7163
Mercator Cylindrical	5367	7924	8444
Mollweide Pseudocylindrical Equal-Area	4052	6068	7118
Peters Equal-Area Cylindrical	3784	5605	7881
Measured with ESRI's ArcView measure tool, approximate values Refer to Figure 4–12			

Table 5–1 Variations in Distance (miles)

Unfortunately, we have to consider the historical nature and quantity of the data. "If you are dealing with data in a specific area, where all the data is in a particular system, you may wish to use a projection that is applicable to that area," said Scott Lobue, a technical specialist for ESRI. Thus, we may be forced to use a map projection for analytical purposes because all of our source data is in this format. "However, if you are working with data that has a wide range of areas, such as the U.S. Gulf Coast, or all of Europe, you may want to keep it in latitudes and longitudes so you can put it in any specific projection at any time."[19]

A different perspective

We can look at ourselves as GIS decision makers, with a whole toolbox of geodetic tools readily available. And we now have an opportunity to break old habits and try to solve problems in a different manner. For instance, are we trying to show the company's assets with a nicely illustrated map or calculate the exact area of a unitized oil field? In essence, solving either one requires a different technique. If the company has assets around the world, for example, a Mercator projection may do the job quite well. But if we wish to calculate area, the ellipsoid is the best computational surface for this purpose. This in turn leads to the following questions:

1. What is the best way to portray the data in a 2-D format?
2. What is the best way to compute distance, area, direction, and volume?
3. How can we set up the geodetic parameters so we can measure, combine, compare, and analyze data in the best manner possible?

In terms of GIS, the first question, implemented through some form of map projection, has everything to do with communicating information. In many cases it has nothing to do with analysis. On the other hand, questions

2 and 3 have everything to do with geospatial computations but in some cases nothing to do with graphical representation. One professional group, the European Petroleum Survey Group (EPSG), even goes so far as to recommend doing away with the grid system altogether when it comes to computing lease areas.

According to the EPSG, "The grid area is unsuitable as a computation of license area because it depends on the choice of projection, and is therefore variable, and because the scale factor inherent in any projection may make a significant difference between the grid and the ellipsoidal areas." EPSG further points out that most GIS applications continue to produce area calculations through grid area calculations. Yet, "Computers have made ellipsoidal area calculations easy and fast and there is now no good reason not to use ellipsoidal area."[20]

An additional advantage of maintaining the GIS project in unprojected geographic coordinates is that it becomes much easier to transform the data to almost any desired projection. Thus, a spherical coordinate system can serve as a hub from and to other projections (see Fig. 5–13).

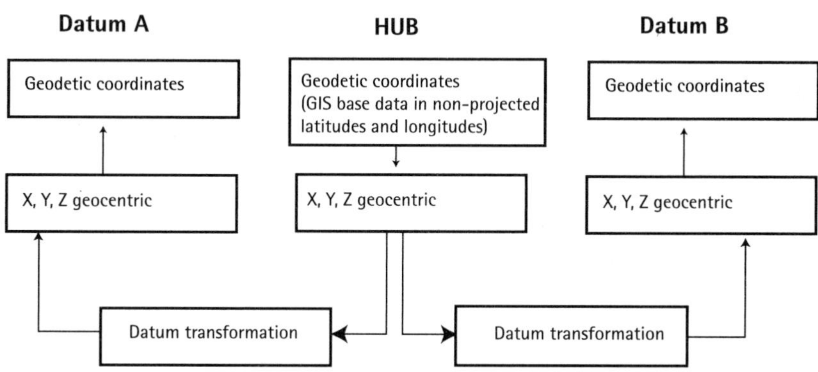

Fig. 5–13 Coordinate Transformation

"And if you have it in latitudes and longitudes, you don't have to worry about the extra baggage such as zone number," says McCabe, "where some of the metadata that you might have to carry are scale factor, Central Meridian, latitude of origin, and so on."[21] If the system begins in a projected X, Y, Z grid, then the deprojection will have to convert the scale factor, central meridian, latitude of origin, and eastings and northings to geographic coordinates (see Fig. 5–14). If the database begins with geographic coordinates, however, this step can be eliminated.

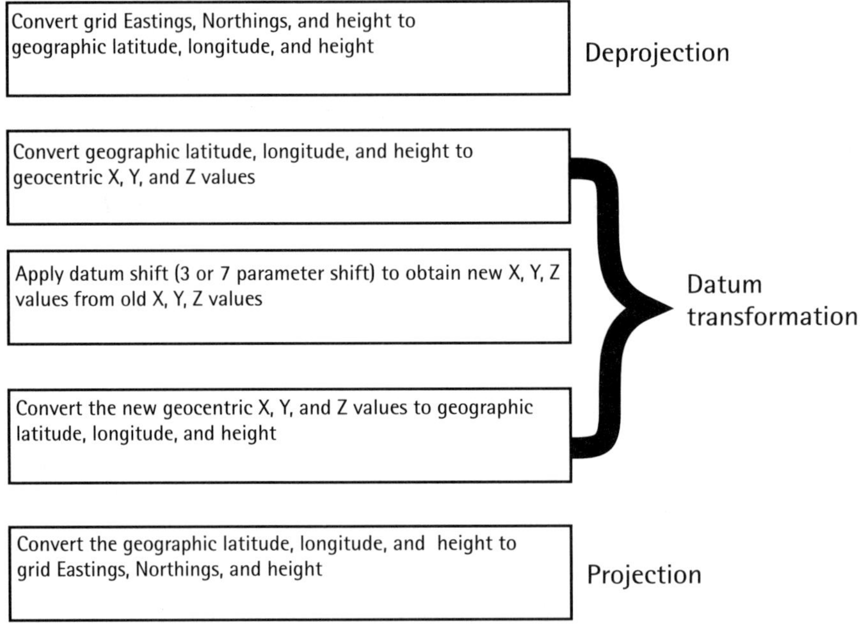

Fig. 5–14 Deprojection

Yet again, we must be careful of software applications. Many programs in the late 20th century (and even more recent programs) did not identify the geodetic system before performing a data import. It was possible to import disparate data sets from NAD27 and NAD83 ellipsoids into the same project and even overlay Cartesian-based satellite imagery into a geographic coordinate system.

"Under some current-generation systems, you need to have your vector data unprojected, your raster data projected, and your project's projection system set to the datum and projection implicit to your raster data," McCabe said. "Otherwise, things won't register properly."[22]

Thus, be aware that some GIS programs will allow you to blindly import disparate ellipsoids, datums, and projections in one system. Fortunately, vendors like ESRI, Intergraph, and MapInfo, with the help of Petrotechnical Open Software Corporation (POSC), EPSG, and other professional groups, are attending to this problem.

Conclusion

Now that we have covered the concepts of geodesy, projections, and grids, you may ask, What does it all mean for the petroleum GIS user? Basically, no other topic affects our work more than geodesy.

First, our database integrity depends on it. It is the job of the petroleum GIS user to know the source data and make sure all geodetic parameters are defined and saved in a unified database. If we use unqualified information, for example, for which the datum remains uncertain, we are certainly headed for trouble. For this very reason, standard geodetic and cartographic data sheets should be filled out and filed for every unconsolidated piece of information (see Fig. 5–15). Additionally, this information should be transferred to metadata files within the GIS. Metadata files are description and reference files containing information on geodesy, data quality, and so forth that are saved with a GIS project. Key notes such as data source, ellipsoid, datum, and projection should be attached to thematic layers.

Country: ______________________________
Project: ______________________________
Digital data format (e.g., SEGPL, UKOOA PI/90, etc.): ______________________________
Geodetic datum: ______________________________
Reference ellipsoid (spheroid): ______________________________
Semi-major axis (a): ______________________________
Reciprocal flattening (1/f): ______________________________
Grid system "x, y" or "Easting, Northing" name): ______________________________
Grid system zone (if applicable): ______________________________
Map projection type: ______________________________
Reference ellipsoid (Must be same as datum for above: ______________________________
Latitude of origin: ______________________________
Longitude of origin or Central Meridian: ______________________________
Scale factor on Central Meridian (Transverse Mercator): ______________________________
Units of Measurement: ______________________________
False Easting: ______________________________
False Northing: ______________________________
For Lambert Conical Conformal or Other Conic Projections: ______________________________
First Standard Parallel: ______________________________
Second Standard Parallel or Scale Factor on First Standard Parallel: ______________________________
If a map is provided with the digital data, what is the Map scale?: ______________________________

Fig. 5–15 Geodetic Data Collection Sheet

Second, geodetic computations allow us to transform and merge data from a variety of sources. Fortunately, GIS applications now allow us to change the map view to whatever projection we want, without changing the underlying data. Unfortunately, geographical calculators and projection engines can be dangerous tools if we do not fully understand the concepts. If users unknowingly commingle entities within the database without ensuring proper transformation of the datums, the database, map images, and all analyses will be damaged. This in turn will produce a ripple effect of location and analytical errors that will affect the entire organization.

Finally, knowledge of geodetic principles, aided by new advances in GIS software, now allows us to optimize the ways in which we portray, measure, and analyze information.

Thus, by obeying basic geodetic and cartographic principles, we can portray, measure, and analyze data easily, accurately, and quickly.

Always keep in mind the following main points:

1. Know the geodetic datum for all data including grids.
2. An ellipsoid by itself does not constitute a datum.
3. To build a paper map, you have to project the data.
4. All projections have distortion—all maps have errors.
5. Computations are best conducted on the ellipsoid.

References

1. Smith, R.S. 1997. *Introduction to Geodesy—The History and Concepts of Modern Geodesy*. New York: Wiley. p. 187.
2. Snyder, J.P. 1987. "Map Projection—A Working Manual." United States Geological Survey Paper 1395. Washington, D.C.: USGS. p. 3.
3. Greenwood, D. 1964. *Mapping*. Chicago, London: The University of Chicago. p. 114.
4. Ibid. p. 126.
5. Snyder, p. 5.
6. Snyder, p. 105.
7. Ibid. pp. 99, 104.
8. Greenwood, p. 139.
9. Greenwood, p. 140.
10. Snyder, p. 12.
11. U.S. Department of the Army. 1981. *Conversion of Warsaw Pact Grids to UTM Grids*. Field Manual No. 34-85. ch. 2, p. 6.
12. University of Wisconsin-Green Bay. Website. (http://www.uwgb.edu/dutchs/FieldMethods/UTMSystem.htm).
13. U.S. Department of the Army, ch. 2, p. 7.
14. Ibid. ch. 2, p. 9.
15. Ibid. ch. 2, p. 11.
16. Rayson, M. Oral Communication.
17. McCabe, M. Oral Communication.
18. Ibid.
19. Lobue, S. Oral Communication.
20. European Petroleum Survey Group. 1995. "Contact Area Definition." *EPSG Guidance Note No.* 3 (November) Petrochemical Open Software Corp. (http://www/ihsenergy.com/?epsg/epsg.html&1).
21. McCabe, M. Oral Communication.
22. Ibid.

6

GPS

One could consider that the ground-based portion of a GPS system and a digitizer are analogous: the Earth's surface is the digitizer tablet, and the GPS receiver antenna plays the part of the cross-hairs, tracing along, for example, a road.

—Michael Kennedy

Global positioning system (GPS) receivers have become the tool of choice for surveying and tracking purposes across our industry. For less than $200, professionals can accurately map surface formations, preplan pipeline corridors, and position drilling rigs using a basic handheld receiver. With the introduction of differential GPS (DGPS), and a significantly larger investment, spatial accuracy can even be fine-tuned to within 1 cm. Finally, GPS data can readily be imported into GIS. This provides a useful tool for tracking the real-time transportation of crude oil and refined products, updating pipeline surveys, monitoring oil spills, or conducting seismic surveys.

If one chooses to use GPS for recreational purposes, technical familiarity with geographic positioning is not of great concern. But for those who use GPS to manage assets, it is essential to understand how it works in order to mitigate or eliminate errors that affect data quality.

This chapter begins by focusing on the three hardware segments of GPS: satellites, receivers, and ground control. The discussion then turns to sources of error, followed by a brief discourse on alternative geopositioning systems. Next, a short explanation on how GPS data can be formatted for transfer into GIS provides some important tips on setup. Finally, unique examples of GIS-GPS integration provide the reader with ideas on how this technology can be applied in the oil industry.

The Hardware

GPS consists of three segments:

1. NAVSTAR constellation of satellites
2. The receiver
3. Ground control[1]

Although the system as a whole is extremely complex, to the end user, its use is rather simple (see Fig. 6–1). Positional accuracies can range as close as just a few centimeters, depending upon several factors. These include: where you are, whether the U.S. military intentionally degrades the transmission, and the level of investment. Most importantly, as long as the user has an unobstructed view of the sky, a position can be taken almost anywhere on the Earth.

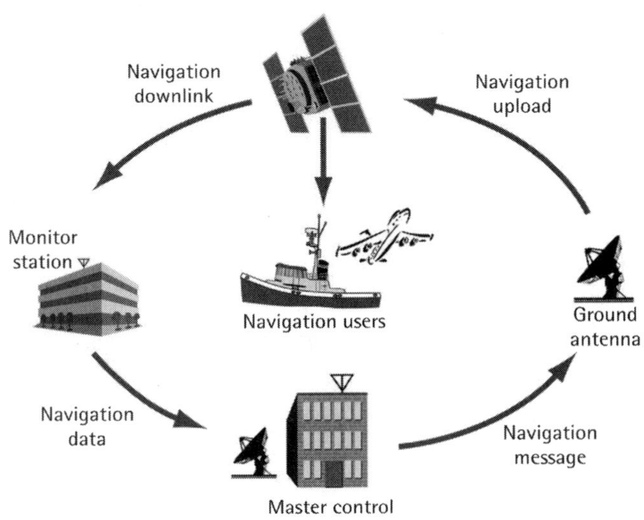

Fig. 6–1 Control Segment and Satellite Constellation (source: courtesy of GPS World.)

Satellites

The GPS satellite constellation (NAVSTAR), developed by the U.S. Department of Defense (DOD), consists of 27 spacecraft (3 spares) orbiting the Earth at a height of 20,200 km. This places the constellation above the standard orbital height of the space shuttle, yet below the geosynchronous orbits of telecommunication and TV satellites.

NAVSTAR satellites reside neither in equatorial or polar orbits, but instead laterally cross the Earth's latitude at an angle of 55°. They execute a single rotation every 11 hours and 58 minutes (12 sidereal hours).[2] In each of six orbital planes, there are 4 satellites, configured so that at least 4 line-of-sight satellites remain in the sky everywhere on the planet. (This is true except in the southernmost and northernmost latitudes.) Each spacecraft weighs about 1900 pounds, uses solar panels for power, and measures 17 ft from tip to tip [see Fig. 6–2 (Plate 6–1)].

The first set of GPS satellites was launched in 1978 (Block I) and the second set in 1989 (Block II). The third set was launched in 1990 (Block IIA) and the fourth in 1997 (Block IIR).[3] Reportedly, the U.S. Air Force plans to deliver more than 30 Block IIF satellites to replace existing GPS Block I, II, and IIA satellites during the next decade.

Ground stations

Satellites must be predictable in four dimensions: three of position and one of time. In order to establish such spatial control, five control stations serve to:

1. Provide exact geodetic "benchmarks" for positional extrapolation from ground station to satellite to target
2. Track current satellite position
3. Predict future position
4. Correct satellite clock time
5. Correct satellite position[4]

The control segment is located at Colorado Springs, Hawaii, Ascension Island, Diego Garcia, and Kwajalein. Except for the master control station at Colorado Springs (Schriever Air Force Base), the remaining ground stations are unmanned and positioned fairly close to the equator. These ground stations constantly receive data from the satellites that are then forwarded to the control station. Control segment positions are exactly defined in reference to the Earth-centered, Earth-fixed WGS84 ellipsoid.[5] As such, they are used as *mile markers* to determine absolute coordinates.

Satellite monitoring and control takes place on a continuous basis with ground stations measuring the distance to each satellite. With time, the orbit of any satellite will degrade. Thus the ground monitor stations keep track of the paths, altitude, location, and speed of each vehicle for correctional purposes. With ephemeris data, future orbital positions for each satellite can be defined. The ground stations send the orbital data to the master control station, which in turn sends corrected data up to the satellite. Ephemeris data are then transmitted as coded information within the navigation message to the GPS receiver.

Although the GPS satellites are free from atmospheric friction, gravitational effects of the Sun and Moon and solar winds influence the orbits as well.[6] Ionospheric and meteorological data sent to the master control station provide some correctional information needed to offset these influences.

Schriever Air Force Base also evaluates and controls repositioning of each satellite's orbit and assesses the overall health status of the satellites. This includes the atomic clocks, internal electronics, power, and booster rockets. This information is then returned to three uplink stations at the Ascension Island, Diego Garcia, and Kwajalein monitor stations. They transmit calibration points to the satellites.

With the older-generation GPS, the DOD could reposition satellites as it did in the 1991 Gulf War. Communications, however, took place through one-on-one communications between each satellite and the ground stations, resulting in a time-consuming procedure. The next-generation GPS, however, will require only a single command from the monitoring station. This command will then be forwarded around the constellation for a quicker and more unified response.[7]

Receiver

Each satellite transmits two low-power radio signals to the receiver (user), designated L_1 and L_2. Civilian GPS uses the L_1 frequency of 1575.42 MHz in the UHF band. [FM radio transmits at 88–108 MHz, often with 100,000 KW of power. GPS ranging signals are broadcast at two L-band radio frequencies: a primary signal at 1575.42 MHz (L_1) and a secondary broadcast at 1227.6 MHz (L_2), using only 50 KW.] These signals can travel through clouds, plastic, and glass, but will not pass through solid objects.

A GPS satellite transmission contains three different bits of information:

1. Pseudorandom code. Identifies which satellite is transmitting information and provides a means of time synchronization between the satellite and receiver
2. Ephemeris data. Contains important information about the operating status of the satellite, current date, future position, and time (www.garmin.com)
3. Almanac. Provides orbital information about the entire system and tells the GPS receiver where each GPS satellite should be at any time throughout the day

As such, the ephemeris and almanac are called the navigation message. The former contains specific satellite information and the latter provides constellation data.

For all of its complexity, GPS ground positioning is determined by simple trilateration. Ranges are based on a simple formula derived from the equation: *distance = velocity x time.* [8]

By recording the speed and time duration it takes a signal to travel from a satellite to a receiver, it is possible to calculate distance. Velocity is taken as the speed of light traveling at about 186,000 mi/sec.

The time differential between satellite transmission and receiver reception, however, is not directly determined. Instead it must be derived by synchronizing atomic clocks within the satellites to a quartz clock in the ground receiver. Obviously, the accuracy of an atomic clock is several orders of magnitude greater than a quartz clock. Consequently it is important to match the two for accurate positioning.

Satellite signals are transmitted in a complicated alternation of binary bits (pseudorandom 0s and 1s) that repeat every 1023 bits. Once the radio signals have been received, the GPS receiver then "peak" matches its own time code with that of the satellite. This is done to calculate the time difference between the two.[9] If there is a time discrepancy, the receiver's quartz clock is automatically adjusted to the time dictated by the satellites. To compute distance, therefore, the GPS receiver must know how long it took for the signal to travel from the satellite. To do this, it takes into account the Earth's rotation in relation to the satellites.

Through trilateration, which requires at least 3 satellites, 3-D positions can be determined by intersecting the distances between the GPS satellites and the receiver.[10] Knowing the distance between a satellite and a receiver, for example, will not pinpoint a location. Instead it presents the user with a nearly infinite number of possibilities that lie along the given circular radius. By adding a second satellite, however, it becomes possible to fine-tune location by reducing the choices to two possibilities along the circular radius. Through the addition of a third satellite, a single point along the radius of the circle can be defined, as well as altitude (in theory).

Finally, the fourth satellite upgrades the correlative accuracy of the atomic clock readings across the satellite-receiver network to establish a high degree of synchronized resolution. Four satellites therefore contribute to finding a point in 4-D space, which in turn allows the user to track stationary and mobile targets.[11] The addition of more satellites refines the overall accuracy of the GPS.

GPS Error Sources

GPS certainly contains a high degree of accuracy as compared to secondary data collection methods. Assume a user digitized a river from a map containing a variable line width 10–15 times larger than the digitizer's cross hair. The accuracy of the digitized points could zigzag from left to right of the centerline of the river by dozens of feet. On the other hand, a GPS, if properly configured, would be able to maintain linear precision from 15 ft to within a matter of inches.

Unfortunately, the user must be aware of several problems that may introduce positional errors. Take the case of a faulty clock. Assume a GPS satellite is moving at a speed of 8653 mph, and the Earth's rotation is moving in some offset direction and velocity. (A point on the Earth's equator, for example, moves eastward at about 1050 mph.) If the onboard clocks or orbital predictions undergo any type of error, ground position will be incorrectly surveyed. A discrepancy of only 1/1000 of a second between the satellite and receiver, for example, could move the user's positional fix 1860 miles away from its true origin.[12]

Some of the other errors that can be introduced include:

- Noise interference resulting in hiccups between the satellite and receiver code
- Selective availability (SA)
- Atmospheric interference
- Human errors

From the introduction of GPS in the late 1970s to the turn of the millennium, the DOD provided civilians with low-resolution measurements. During this time, the DOD relayed high-resolution data only to the military.[13] In the civilian system, the DOD intentionally manipulated clock frequency so as to spread out the distribution of positional measurements. In this case, the government guaranteed that a fix would remain within 100 m of the true position 95% of the time. The DOD originally designed GPS as a weapons support system. As a result, the military obviously did not want to risk giving the enemy access to a high-resolution geographic positioning system.

In May 2000, however, former President Clinton made good on his promise to untether the civilian restriction and allow commercial and personal use of unadulterated signal generation. Because of this, out-of-the-box GPS hand receivers now have an accuracy of 10–15 m. Nevertheless, the military still retains the option of shutting the system down or reintroducing SA in times of emergency.

Other sources of error concern the tropospheric and ionospheric delays associated with changes in temperature, pressure, humidity, and ionized air. (The troposphere is located 8–13 km above sea level. The ionosphere is located 50–500 km above sea level.) Multipath errors also occur when signals are reflected from surfaces near the receiver such as water or sandy soil. These reflections interfere with the signal between the satellite and the receiver.[14]

Human errors

GPS receivers can be dangerous tools in the wrong hands if users are not educated in their use. There are many documented incidents where the simplest of misunderstandings resulted in dramatic errors in the horizontal position of target features.[15] Assume a pipeline engineer provides a pipeline corridor in a UTM grid coordinate without converting to an Earth-centered, Earth-fixed system (WGS84). If operating personnel begin laying a pipeline with a GPS receiver, but forget to convert planar coordinates to spherical, obvious problems will develop.

Human mistakes can by themselves result in errors of several kilometers. Referring to chapters 4 and 5, it is always best to consult with a geodesist before inserting data into a GIS that has a different ellipsoid than WGS84. Although many GPS receivers will display different coordinate systems, the data that is downloaded will most likely remain in WGS84.

Finally, GPS requires a direct line of sight to the NAVSTAR satellites. Thus, it does not work in forested areas or in buildings, underground, or when it is heavily raining. Some experts advise as many as 5 line-of-sight communication satellites before assuming a high degree of confidence in the coordinates. As such, it becomes important to evaluate the geometry and number of satellites in the sky. This is known as positional dilution of precision (PDOP).[16]

Ideal satellite geometry exists when the satellites are located at wide angles relative to each other.[17] Poor geometry results when the satellites are located in a line or in clusters. Some of the more advanced GPS units allow the user to forecast sky positions in order to ascertain the best times for data collection.

Altitude errors

Using GPS to determine elevation can also be confusing for an unwary user. On a visit to a jackup in the Gulf of Mexico, for example, this author used a handheld GPS to spot the well. The device, however, recorded the height of the helicopter pad at 124 ft, where it should have been 102 ft (as tape-rule measured by rig personnel during midtide).

This error occurred because GPS computes height relative to the WGS84 ellipsoid, a computational surface that best represents the geoidal surface globally (see chapter 4). The height determined by the satellite receiver, in this case, is called ellipsoidal height or geoidal height. *This elevation must not be confused with orthometric height*, which is referenced to mean sea level. Due to the geodetic principles that GPS abides by, these devices cannot readily provide orthometric height for ready comparison with topographic maps.

Only when the geoid and ellipsoid coincide will the GPS read correct height.[18] Otherwise, these surfaces will have some distance of negative or positive separation. To turn GPS ellipsoid values into their equivalent orthometric value (true height above sea level), a correction factor must be applied.[19] This geoidal separation can be found with varying degrees of accuracy. Where observational coverage is well developed, contour maps can be interpolated for the target location. The Internet provides a useful source for finding these maps. Some services also provide computer models to predict values.

Differential GPS

The errors introduced through SA, clock errors, atmospheric delays, and so forth have been resolved with the introduction of DGPS. With DGPS, the differential process consists of setting up a GPS receiver/transmitter, called a base station, at a precisely known geographic point (see Fig. 6–3). Because the base station's exact position is always known, it can be used to analyze and record faulty GPS signals received from the satellites.[20] The base station and rover are placed in the same atmospheric corridor, 300 mi or less from each other. Consequently, it can be assumed that both will be subject to the same magnitude of error. Hence, the base station's exact fix can be used to differentially correct the user's (or rover's) geographic position.[21]

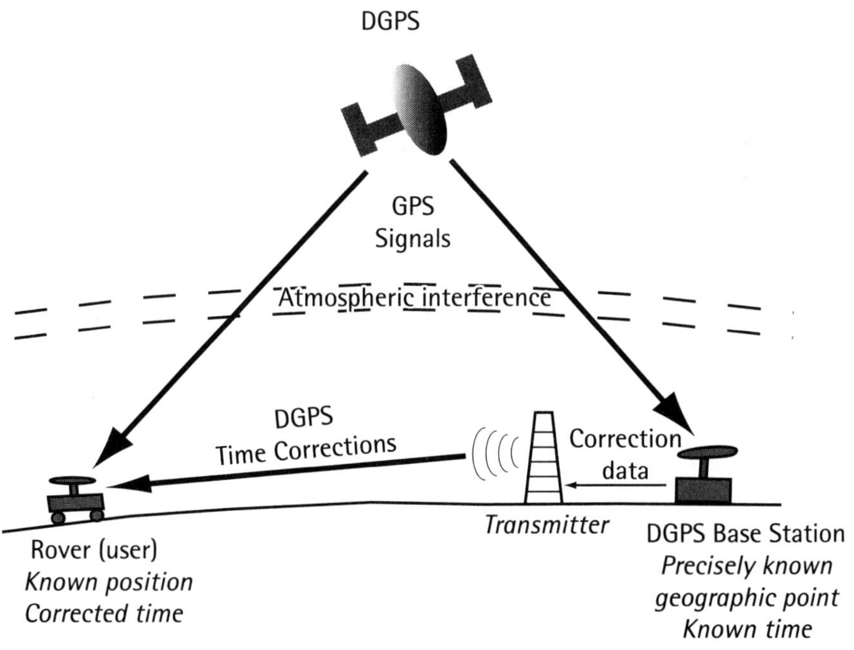

Fig. 6–3 Differential GPS

The basic process works through reverse calculations through the prediction of total signal delay. In this case, the stationary receiver uses its known position to calculate timing, rather than using timing signals to calculate position (as it does with normal GPS). It can then derive what the correct travel time should be and compare this value to the actual value received. The difference in time then becomes the correction factor, which is subtracted from the measurements obtained from the roaming receiver. This correction is applied to every satellite signal reading.

The sum of the information is then communicated through a wireless data link to the roaming receivers. Generally, if the rover remains within 100 mi of a base station, the resulting positions will be accurate to within the technical specifications of the receiver. The farther away the user moves, however, the less accurate the position becomes.

Some DGPS users, such as those on deepwater drilling ships, utilize real-time DGPS processing to maintain station. This is because the vessel is affected by wind, waves, and currents that move the drillship away from the center line of the well. In many cases, the drilling rig integrates the DGPS signals with onboard computers. Thrusters are initiated in whatever direction needed to keep the vessel's rotary table directly above the wellbore.[22] Tankers and pipeline lay vessels use DGPS as well to navigate or work in coastal waters.

Correctional data can be free, except for the cost of a receiver, if the user is located near a public base station. For example, the U.S. Coast Guard has placed several beacons in coastal areas of the United States. In other cases, industries must pay for and build their own system. Reported costs of such systems are in the range of $50,000. (See Box 6–1 for a discussion of other systems.)

Box 6–1
Other Systems

The Russian Federation's counterpart to GPS is GLOSNASS. GLOSNASS provides 100-m accuracy with SA deliberately degraded (as formerly with GPS) and 10–20-m accuracy for military signals. Conceived and promoted in the early 1970s by the former Soviet Ministry of Defense and the Soviet Navy, the original design by the Applied Mechanics NPO and the Institute for Space Device Engineering depicted 21 satellites in three orbital planes at an inclination of 65° with three spares.

First deployment took place October 12, 1982 with satellites orbiting the Earth in nearly circular paths at an altitude of 11,900 mi. The latest GLOSNASS launch took place December 1, 2001 with 7 operating satellites in late 2002 (August 1, 2002, GPS World, Almanac). Reportedly, positional data for oilfield operations have been made available to users seeking primary data redundancy for GPS.

The Europeans also intend to develop a satellite-based positional system, called Galileo, that would directly compete with GPS. Initially targeted for deployment in December 2000, political pressure from the United States and budget problems in the European Union have delayed operational development. Proponents of the system, however, continue to move towards putting the system in place (Galileo's World, Oct. 25, 2002). Unfortunately, international demand for radio frequencies, especially from China, has placed pressure on European developers to get it up and running. Otherwise the requested frequencies will be taken.

Because Galileo will be run through a public-private partnership, rather than exclusively through the military, its success will be based upon commercial usage rather than government support. It is hoped that Galileo will be fully functional by 2008. Other issues to be resolved include signal compatibility with GPS and GLOSNASS, potential integration, and most importantly, world security.

Data Integration

GPS receivers can be set up to record current and historic location data into memory. Thus GPS-derived coordinates tied to entity type and attribute description can be directly exported into GIS, often through a serial port connection. Hands-on operation and integration with GIS nevertheless require preparation and planning.

First the user has to determine what kind of system is preferable (handheld vs. DGPS). Then the designer must format the way in which data are exported from the GPS into GIS. For this to work, the user must define a GPS data dictionary that is then written into the receiver's software package. To create a data dictionary, every entity must be categorized as a point, line, or polygon (see chapter 3). Additionally, the necessary attributes needed to describe these entities must be custom defined. For a seamless transfer of data, the GPS's data dictionary must therefore exactly match the field attributes and entity definitions in the GIS.[23]

There are several tricks of the trade for setting up a data dictionary. Try, for example, to think about an entity that requires two or more words, such as *oil well*. In this case, the designer of the data dictionary should exclude the space in between both words or insert an underbar. Otherwise, the GIS database may recognize *oil* and *well* as two separate items.

Additionally, attribute names must be distinguishable from one another. The user might capture, for example, the type of a particular asset, such as an artificial lift pump, but fail to distinguish it from another asset, such as a pipeline pump. If this takes place, then the GPS program may merge all of the assets into one file upon GIS export. To avoid this, the designer should incorporate a short prefix that corresponds to the asset such as: *RodP_Size* and *Centrif_Type*.

Examples

GIS and GPS integration are inherently compatible. In fact, there are so many unique applications that this topic deserves its own book. (See Kennedy for an excellent guide on both theory and field application.) One firm, for example, tracks the locations of bar-coded pipe using a GPS receiver. It then exports the data to a GIS for planning and maintenance purposes over the lifetime of the pipeline.[24] "The electronic pipe data captured in the field is then quality assured against the digital mill manifests for abnormalities or identification number errors." The mill manifests are then used within a handheld computer for on-the-fly quality assurance as pipe data are being collected.

Rig tracking

One extremely interesting application includes real-time GPS-based drill rig tracking via GIS. The system consists of three basic parts: a GPS receiver mounted to the rig, two-way satellite communications, and an Internet-based GIS (www.spudit.com/spudit/).

The system works as follows:

1. Rig-based GPS transponder sends a signal to one of the geostationary satellites (uplink), initiating a coordinate upload that eventually tells the end user where the rig is located
2. Geostationary satellite forwards signal to an Earth station (downlink—not to be confused with a GPS base station)
3. Earth station digitizes the signal and sends the data to the company's global rig database via the Internet
4. Updated database becomes available for retrieval via web-based GIS
5. User logs onto the website and activates rig database through the GIS software program (database activation)
6. User at this point can either search or query the database for rig attributes or produce maps

According to Keith Fraley, GIS manager with SpudIT LLC, the company applies a three-tier ESRI GIS web model as follows:

- Oracle 8i with ArcSDE back end
  - o ArcIMS middle-tier application server
  - o ArcIMS/ASP front-end user interface
  - o XML-based communiqué between tiers
- Routing, geocoding, and reverse geocoding ability
- Easy-to-use query interface
  - o Rig attribute query ability
  - o Spatial/geocentric query functionality[25]

In this example, integrating the rig-based GPS receivers with an Internet-based GIS has provided some beneficial utilities. Operators, contractors, suppliers, and service companies, for instance, can locate available rigs [see Fig. 6–4 (Plate 6–2)], query rig equipment specifications, and even obtain travel directions to the wellsite. As Figure 6–4 (Plate 6–2) shows, drilling rigs mounted with GPS sensors can be tracked in real time no matter where they may be. In this web-based application, users can employ many of the search, query, and analytical tools available to common GIS programs such as buffers and network analysis.

According to Robert G. Davis, president of SpudIT LLC, about 1100 North American rigs have been mounted with GPS transponders. The database for many of these rigs contains attribute data on drawworks size, pump capacity, drillstring components, blowout preventers, and so forth. With this type of data, many of the GIS search and analysis capabilities discussed in chapters 2 and 3 can be brought into play.

"If the drilling manager needs to locate all available 2000-hp rigs within 200 miles of a prospect," Davis explained, "he can conduct a search for cold-stacked, hot-stacked, recently released, or immediately available rigs. Thus, the user can utilize a buffer-based query." Furthermore, "A user can calculate mileage and 'mob' costs by analyzing the various routes from each available rig to the prospect for economic comparison with one another."[26] Therefore, network techniques such as shortest path analysis can be used (see chapter 3).

Service companies and hot-shot services appear to be the most active users of the system. They take advantage of detailed driving instructions and maps that save time in finding the rigs with hard-to-find locations.

Conclusion

GPS technology can be used to gather geographic and attribute data directly from stationary and moving targets. Primary data collection like that gathered from GPS has several advantages over indirect methods, such as digitizing. These advantages include improved accuracy and the ability to monitor objects in real time. For GIS users, the data can be easily imported into GIS, as long as the GPS data dictionary matches the database fields in the GIS. Great care must be taken not to confuse coordinate systems and recorded height using GPS. Otherwise, errors will be inserted into the database. As GPS technology becomes more widely accepted, the number of collecting stations and the amount of data will grow. This will, in turn, provide an expanding network of inputs. As such, the overall quality of data collected across the petroleum industry will improve.

References

1. Dana, P.H. "The Geographer's Craft Project." *Global Positioning System Overview*. University of Colorado-Boulder (online).
2. Kennedy, M. 1996. *The Global Positioning System and GIS: An Introduction*. Chelsea, MI: Ann Arbor Press. p. 11.
3. *GPS World Almanac.*
4. Parkinson, B.W. and J.J. Spilker, eds. 1996. *Global Positioning System: Theory and Applications*. Vol. 1. Washington, D.C.: American Institute of Aeronautics and Astronautics. p. 29.
5. Parkinson, pp. 439–440.
6. Kennedy, p. 13.
7. Steede-Terry, K. 2000. *Integrating GIS and the Global Positioning System*. ESRI. p. 12.
8. Ibid. p. 8.
9. Ibid.
10. Rayson, M. 1999. *GPS InQuest*. Ver. 2.00 (CD ROM) Quality Engineering and Survey Technology Ltd.
11. Kennedy, p. 57.
12. Steede-Terry, p. 7.
13. Kennedy, p. 124.
14. Parkinson, p. 52.
15. Rayson, M. Personal communications.
16. Steede-Terry, p. 26.
17. *Garmin GPS Guide for Beginners*. 2000. Website tutorial at www.garmin.com.
18. Rayson, M. Oral Communications.
19. Smith, R.S. 1997. *An Introduction to Geodesy—The History and Concepts of Modern Geodesy*. New York: Wiley & Sons. p. 75.
20. Kennedy, p. 120.
21. Steede-Terry, p. 14.
22. "Drillships Target Deepwater Operations." 1999. *Oil & Gas Journal* 97:34 (August 23) p. 70.
23. Steede-Terry, p. 24.
24. Pryor, R. 2001. "GIS for Oil & Gas Conference Proceedings." *Tenth International Conference and Exposition, GITA*. Houston (September 17–19). p. 307.
25. Fraley, K. Oral Communication.
26. Davis, R. Oral Communication.

7

Remote Sensing

Remote sensing refers to those methods that employ electromagnetic (EM) energy, such as light, heat, and radio waves to detect and measure target characteristics.[1] Remote-sensing data are intrinsically related to some position in space. As such, it is arguably one of the most valuable sources of data for GIS. In fact, the development of GIS and modern remote sensing coincided, each with beginnings dating back to the 1960s.[2]

Geoscientists have long applied photographic cameras, radar, lasers, infrared (IR) scanners, radiometers, spectrometers, microwaves, and multispectral scanners (MSS) in the search for hydrocarbons. Many of the prolific Wyoming oil fields associated with tectonic folding, for example, were clearly identified by aerial photos and stereoscopes in the 1950s and 1960s. With the introduction of satellite remote sensing, these basic techniques were then coupled with new technologies. This produced enhanced views of the Earth's surface.

In Figure 7–1 (Plate 7–1), for example, a synthetic stereo pair was created by merging a TM 2-3-4 image with digital elevation data for Sheep Mountain, Wyoming. Intrinsically, Landsat images have no stereo capability. Thus synthetic stereo models are produced by digitizing topographic contour maps in a raster array of pixels.

Today, it is not uncommon for a geologist interested in identifying a prospective area to begin with a small-scale Landsat satellite image of the basin. The next step is to overlay pertinent magnetic and gravity surveys, narrowing in on the area of interest with larger-scale high-altitude aerial photos. Vector-based data, such as wells and oilfield infrastructure, are then emplaced within. The geologist may then complement this data with vector-based structure contour maps, often converted to a raster format that has been tied to well and seismic information.

Remote sensing adds its greatest value by representing reality through the expansion of our senses. In turn, traditional map views can be enhanced with data that extend beyond the visible spectrum.

Like the prior subject on GPS, remote sensing cannot be described with a single chapter. For further study, the author advises several specific titles. These include: *Remote Sensing: Principles and Interpretation*, Sabins; *Remote Sensing for Geologists*, Prost; and *Raster Imagery in Geographic Information Systems*, Morain and Baros.

To introduce the subject, this chapter begins by focusing on the EM spectrum, an important topic that must be understood before beginning to work with remote-sensing data. The discussion then turns to aerial photography and satellite imagery, two key areas of data acquisition. Finally, examples of remote-sensing GIS integration provide the reader with historic representations on how this technology has been applied in the petroleum industry.

The Visible and the Invisible

Humans are blind to all wavelengths except a narrow band of colors ranging from blue to red (see Fig. 7–2). Because this portion of the spectrum controls the way in which we see the world, we tend to use this slice of light as our main guide to the world around us. Obviously this leaves an entire universe out there that would remain invisible without the help of EM sensors.

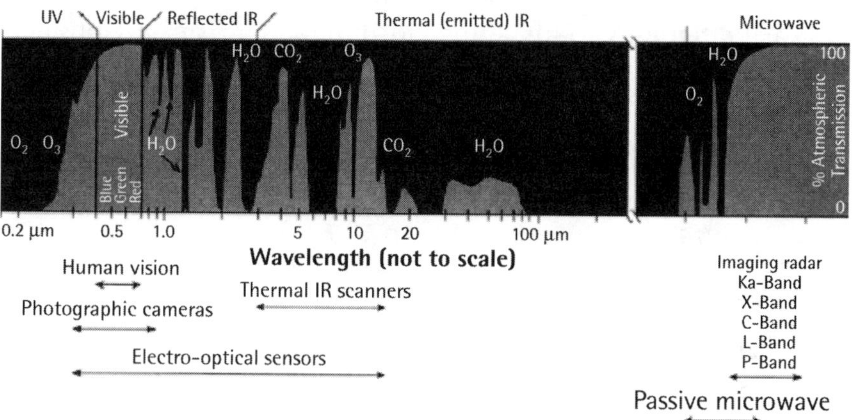

Fig. 7–2 The Electromagnetic Spectrum (source: NASA)[4]

To understand how the EM spectrum works, think of waves and ripples as they travel along a body of water. You can directly see and feel these by body surfing or riding a raft, or even make new patterns by swimming or paddling. But there are other waves in the spectrum that cannot be seen, although at times they may be felt (heat) or heard (sound). One particular group of waves, called EM waves, can pass through any medium including space, air, liquids, and solids. (Because sound requires some sort of molecular medium through which to travel, it is not considered part of the EM spectrum.) This class of data provides the majority of remotely acquired data.

EM spectrum

All surfaces and objects such as outcrops, soils, vegetation, and artificial structures reflect and emit energy throughout the EM spectrum. Each feature contains unique characteristics. From these, various investigations by EM sensors can help differentiate physical, optical, and chemical features. Unconsolidated silicate sand deposited along the strandline of a shore, for example, will react in a certain way to an EM probe. An outcrop of well-cemented, feldspar-rich sandstone will react differently. This ability to differentiate various physical properties makes it possible to measure, map, and employ imagery throughout time. Thus many phenomena such as weather and urban sprawl can be tracked.[5]

The EM spectrum can be described as a continuum of energy whose wave form:

- Measures from nanometers to meters in length
- Travels at the speed of light
- Propogates through a vacuum such as outer space[6]

The EM spectrum is divided into regions such as gamma ray, X-ray, ultraviolet (UV), visible light, IR, microwave, radar, and radio (see Table 7–1). Because the Earth's atmosphere absorbs energy in the gamma-ray, X-ray, and UV regions, these regions are not traditionally used for remote sensing. As such, industrial use of remote sensing takes place mostly in the long wavelength UV, visible, IR, and microwave regions.

UV WAVELENGTHS. UV wavelengths measure 0.25 micrometers (μm) to about 0.40 μm in length. They are shorter than visible blue and are mainly applied in fluorescence studies for mineral exploration and oil spill detection. Although this band can be detected with film and photo detectors, atmospheric interference can be severe. Wavelengths shorter than 0.3 μm are completely absorbed by the ozone layer in the upper atmosphere.[7]

Wavelength Region	Wavelength	Comments
X-ray	< 0.03 nm	Entirely absorbed by the Earth's atmosphere. Unavailable for remote sensing.
Gamma Ray	0.03–30 nm	Entirely absorbed by the Earth's atmosphere. Unavailable for remote sensing.
Ultraviolet	0.03–0.4 μm	Wavelengths from 0.03–0.3 μm absorbed by ozone in the Earth's atmosphere.
Photographic Ultraviolet	0.3–0.4 μm	Available for remote sensing. Can be imaged with photographic film.
Visible	0.4–0.7 μm	Available for remote sensing. Can be imaged with photographic film.
Infrared	0.7–1000 μm	Available for remote sensing. Can be imaged with photographic film.
Reflected Infrared	0.7–3.0 μm	Available for remote sensing. Near Infrared 0.7–0.9 μm. Can be imaged with photographic film.
Thermal Infrared	3.0–14.0 μm	Available for remote sensing the Earth. This wavelength cannot be captured with photographic film. Instead, mechanical sensors are used to image this wavelength band.
Microwave or Radar	0.1–100 cm	Longer wavelengths of this band can pass through clouds, fog, and rain. Images using this band can be made with sensors that actively emit microwaves.
Radio	0.1–100 cm	Not normally used for remote sensing the Earth.

Table 7–1 The Electromagnetic Region (source: courtesy M.J. Pidwirny, Dept. of Geography, On-line, Okanagan University College)

VISIBLE REGION. The visible region extends from about 0.40 µm (blue) to about 0.70 µm (red). This region records pigments resulting from particular chemical and molecular structures that are typically captured with film.[8] Both aerial photography and satellite imagery collect data from this region. The visible region has been used by the petroleum industry for a variety of reasons, such as stereographic mapping of surface formations.

IR. IR radiation occupies the region between 0.70 µm and 3.0 µm. This band is used mainly as a means for distinguishing vegetation. Healthy green vegetation, for example, induces a very strong reflector of IR radiation that appears bright red on color IR photographs.

SHORT WAVELENGTH. At wavelengths between 1.5 µm and about 3.0 µm, called the *short wavelength* region, detected energy consists of a mixture of reflected and emitted radiation.

These wavelengths are mostly applied in atmospheric studies and can be used to distinguish snow from clouds.

THERMAL IR BAND. From 3.0–14 µm, called the *thermal IR* band, very hot objects can be detected. These include volcanoes, power plants, industrial discharge waters, and forest fires. The main difference between thermal IR and IR is that thermal IR is *emitted* energy, whereas the near IR (photographic IR) is *reflected* energy. At these wavelengths, images are acquired by optical-mechanical scanners and special vidicon systems, but not film.[9] It should be noted that the peak power radiation of most Earth objects occurs between 8.0 µm and 14.0 µm.[10]

Microwave. Moving into the millimeter wavelengths, *active microwave* (radar) and *passive microwave* (radiometry) come into play. These longer wavelengths can penetrate clouds, rain, and fog but are sensitive to dielectric properties of materials and moisture content. Radar is also sensitive to object geometry. One interesting application of radar imaging came into play in 1996 when Russian and British scientists integrated ice-penetrating radar, spaceborne altimetric (ERS-1), and downhole seismic data. These were combined in an effort to determine the horizontal extent of a subterranean Antarctic lake (Lake Vostok). Located 3.6 km below the ice sheet, it turns out that the areal and volumetric sizes of this lake are larger than Lake Ontario [see Fig. 7–3 (Plate 7–2)].

Many of the technologies used in these scientific investigations were adapted from the petroleum industry.

Interaction Processes

Remote sensor systems can be broadly categorized into active and passive systems. Passive remote-sensing systems record energy that naturally radiates or reflects from an object. A camera is an example of a passive system. It records the light energy that is illuminated from the Sun. Add a flash attachment to the camera, however, and it becomes an active system. In a similar way, active microwave and radar systems provide externally sourced illumination by transmitting a signal. They then receive an echo from the ground or object.

Most satellite remote-sensing data are currently acquired by MSSs and linear array devices. These are passive systems that record solar radiation reflected from the Earth's surface. Data derived from MSSs can provide information on vegetation types, geomorphology, soils, surface waters, and river networks.[11]

EM energy that interacts with matter can change the intensity, wavelength, direction, polarization, and phase of incident radiation.[12] Because EM sensors can record these changes, investigators can interpret the various *unseen* properties of matter.

Incident radiation and its associated energy may be:

1. Transmitted, or passed through a substance. The transmission of energy through media of different densities, such as from air into water, causes a change in the velocity of EM radiation. By comparing the velocity of energy within a vacuum to the velocity in the substance, specific properties of the matter can be studied.
2. Absorbed, or released largely due to the heating of the matter.
3. Emitted, or discharged from the substance as a function of structure and temperature.
4. Scattered, or deflected in all directions. Surfaces with dimensions of relief produce scattering.
5. Reflected, or returned from the surface of a material with the angle of reflection equal and opposite to the angle of incidence. Reflectance is caused by surfaces that are smooth relative to the wavelength of incident energy. Polarization, which is functionally related to the direction of vibration of the reflected waves, may differ from that of the incident wave.

These interactions can be broken down into *surface* or *volume-based phenomena*. *Surface-based phenomena* depend on the exterior properties of the object. They are usually a function of color or roughness and result in the emission, scattering, or reflection of EM energy. *Volume-based phenomena* depend on the interior properties of the object. They are usually a function of density and conductivity and result in the transmission or absorption of EM energy.

The particular combination of surface and volume interactions with any particular material depends on both the wavelength of the EM radiation and the specific properties of that material.[13] Interactions between matter and energy can be recorded with remote sensing, after which conclusions can be drawn as to the characteristics of matter.

Aerial Photography

The simplest and earliest forms of remote sensing used photographic cameras to record information from visible wavelengths (see Table 7–1). This era began in 1858 when photographer Gaspard Felix Tournachon equipped a balloon with a camera near Paris, France. In the following decade, the role of balloon-based aerial photography expanded as the military used it to gather information about enemy troop movements during the U.S. Civil War.

In 1909, Wilbur Wright became the first person to take aerial photos from an airplane. This was soon followed by civilian use, with airplane-mounted aerial photography acquiring systematic vertical images across Canada, the United States, and Europe.[14] In the petroleum industry, one notable entrepreneur was Edgar Tobin. He started a company in 1927 to supply aerial photographs to oil and gas companies operating in the Gulf Coast states of the United States. His efforts led to the basis for creating maps of entire counties that set cartographic standards in North America.

It must be understood that aerial photos of the Earth's surface are not maps, but instead are single-point perspective views (see Fig. 7–4). As such, they contain geometric distortions that depend on the focal length (f) of the lens and the flying height above the ground (H). Because flying height varies with topography, for example, scale is not constant throughout a photo. This relation between f and H

can therefore be used to determine photo scale (PS) by applying the formula $PS = f \div H$. The art of mitigating distortion in aerial photography led to the science of photogrammetry.

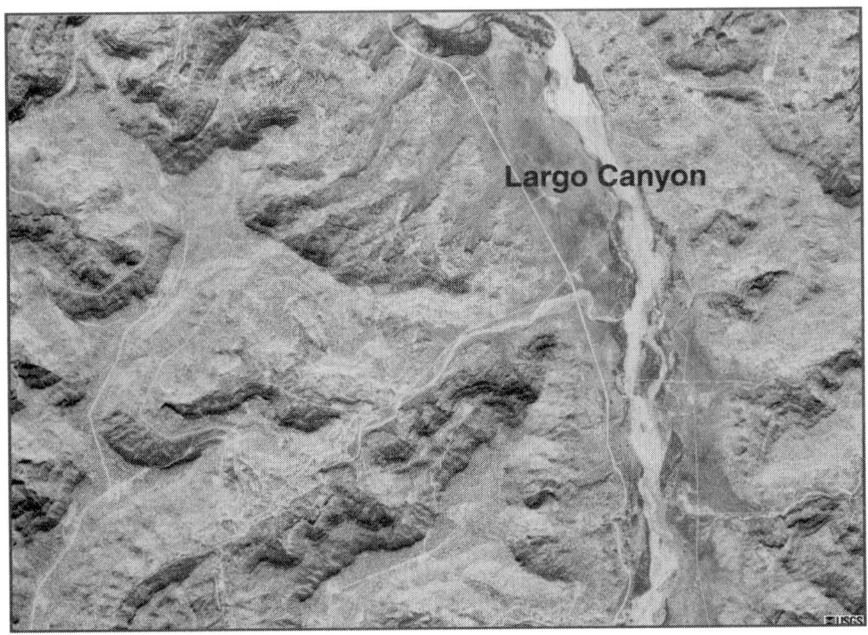

Fig. 7–4 Aerial Photograph of Largo Canyon, NM (source: USGS)

The aerial photograph shown in Figure 7–4 can be compared with the topographic map shown in Figure 2–1. Note that such features as roads and surficial expressions are more easily depicted with the former. It must be remembered, though, that aerial images cannot be used in GIS until they have been orthorectified and identified with a coordinate system.

Photogrammetry uses height differences, absolute elevations, and the dimensions and locations of objects to determine accurate spatial representations. It should be mentioned that most topographic maps are still based on air photos.

Aerial photos are acquired by aircraft using cameras specially adapted so that the film advance and the shutter exposure are synchronized with flight speed.[15] This ensures that each frame has at least 60% forward overlap and 30% sidelap, traditional criteria for stereographic viewing. (Special equipment utilizing a stereoscope allows specialists to view overlapping photos in 3-D.) For geologic applications, aerial photography is usually acquired between midmorning and midafternoon. During this time frame, the Sun is high and shadows have minimal effect.

For use with GIS, it is extremely important to correct the problem of relief displacement, a geometric distortion that is present on all vertical aerial photographs.[16] When one looks at the tops of buildings and cliffs using aerial photographs, such objects appear to lean away from the optical center of the photograph. As displacement increases from the photographic center to each of the photograph's corners, this condition worsens.

Yet the amount of relief for an aerial photograph has several useful attributes. It is:

- Directly proportional to the height of the object
- Directly proportional to the radial distance from the principal point to the top point corresponding to the top of the object[17]
- Inversely proportional to the height of the camera above the ground

These relationships can be used to determine the height of an object from its relief displacement.

Consider the diagram shown in Figure 7–5. When an airplane-mounted camera captures an image of the Earth's surface, features become radially displaced outward from their true planimetric positions due to lens convergence. If an unrectified image is then used to calculate distance between two or more points, such as A'–B' instead of 1–6, measurements will be underestimated.

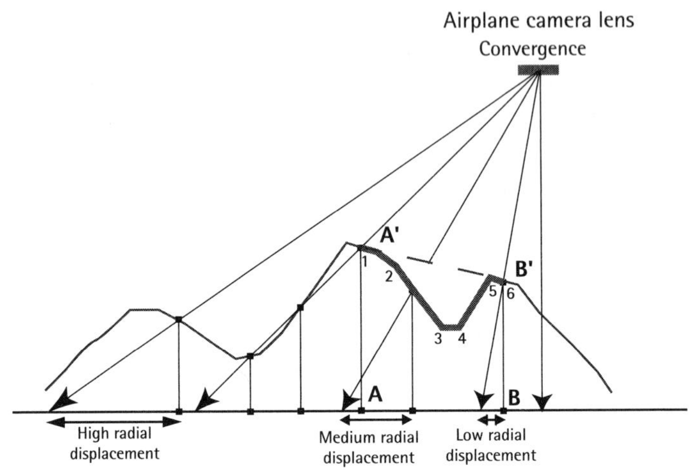

Fig. 7–5 Distortions in Aerial Photographs

Obviously, if such distortions are not removed, aerial photos cannot be used in GIS. Once resolved into orthophotographs, however, they can be readily used in GIS. Orthophotographs are a thematic data set that has been scanned into digital format and computer-processed to remove distortion. Orthorectified data maintain a consistent scale throughout the image (see Fig. 7–5) and as such can be used much like a map.

In fact, its closer representation of Earth's surface, as compared to contour maps produced through human interpretation, makes orthophotography the preferred medium for GIS analysis. It is especially useful as it can be overlaid with other themes.[18]

Satellite Remote Sensing

Aerial photography still provides the majority of remote-sensing data used today. However, the wide variety of EM sensors installed on space-borne vehicles shows significant growth in data acquisition.

It has greatly expanded our view of the Earth, ranging from atmospheric studies to land usage. Most of these data are available in one of two formats: *digital* (electronic or vector) and *analog* (image or raster).

From the very beginnings, satellite remote sensing leapt into the digital age. The first meteorological satellite TIROS-1 (April 1960), for example, did not utilize conventional film. Instead it captured images digitally then transmitted this information to receiving stations on the Earth. Unfortunately, scientists and users were not able to take full advantage of digital functionality until the computer age in the 1980s.

Landsat (United States)

In the 1970s, civilian remote sensing received a significant boost with first deployment of the Landsat satellites. Since this time, several generations of Landsat satellites using MSS scanners have provided increasingly useful images of the Earth's surface. Landsats 2 and 3 were launched in 1975 and 1978, respectively. They collected simultaneously reflected and emitted data over 185-mi wide swaths from four bands of the EM spectrum, ranging from visible green through near IR:

- Band 4, 0.5–0.6 μm (green)
- Band 5, 0.6–0.7 μm (red)
- Band 6, 0.7–0.8 μm (near IR)
- Band 7, 0.8–1.1 μm (near IR)[19]

Thematic mapper (TM) sensors were installed on Landsats 4 and 5 in 1982 and 1984 [see Fig. 7–6 (Plate 7–3)]. The bandwidths were then expanded beyond MSS to the far IR portions of the EM spectrum for a total of seven bands, with resolutions on the order of 30 x 30 m.[20]

Landsat 6 never made it into orbit. However, the launch of Landsat 7 on April 15, 1999, began a new era with the addition of the

Enhanced Thematic Mapper Plus (ETM+) scanner. The ETM+ is a fixed-position, nadir-viewing, whiskbroom, multispectral scanning radiometer. (A whiskbroom scanner uses a telescope directed towards a mirror that is oriented at 45°, rotating back and forth about an axis parallel to the line of flight.) It is capable of detecting eight bands at a resolution of 28.5 m (see Table 7–2).

Satellite	Instrument	Resolution (m)	Coverage (km)	Spectral Bands	Stereo	Min. Revisit (days)	Archive (years)	Date Launched (Proposed Launch)
IKONOS-2 USA	Panchromatic Multispectral	1 4	Custom	1 4	Yes Yes	3 3	3	Sept. 1999
QuickBird-2	Panchromatic Multispectral	0.6 2.5	Custom	1 4	Yes Yes	5 5	1	Oct. 2001
EROS-A1	Panchromatic	1.8	12.5 x 12.5	1	Yes	5	2	Dec. 2000
SPIN-2 Russia	Panchromatic Panchromatic	2 10	40 x 180 200 x 300	1 1	No Yes	45+ 45+	18++	Feb. 1998
IRS-1C/D India	Panchromatic Multispectral	5 23.5	70 x 70 & Custom 141 x 141	1 4	No No	5 5	6	Dec. 1995 / Sept. 1997
SPOT-4 France	Panchromatic Multispectral	10 20	60 x 60 60 x 60	1 3/4	Yes Yes	5 5	4	March 1998
SPOT-5 France	Panchromatic Multispectral	2.5 and 5 10 and 20	60 x 60 & Custom 60 x 60 & Custom	1 4	Yes Yes	5 5	–	May 2002
LANDSAT-7 USA	Panchromatic Thematic Mapper	15 28.5	185 x 170 185 x 170	1 7	No No	17 17	3	April 1999
Radarsat-1 Canada	Radar C band / HH	10–100	50 x 50 to 500 x 500	1	Yes	24	7	Nov. 1995
OrbView-3 USA	Panchromatic Multispectral	1 4	Custom	1 4	Yes Yes	3 3	–	(Q4 2003)
ALOS Japan	Panchromatic Multispectral Radar - L Band Polarimetric	2.5 10 7–100	70 x 70 70 x 70 20 x 20 to 350 x 350	1 4 V & H	Yes Yes –	2 2 2	– – –	(Q1 2004)
EROS-B1 USA/Israel	Panchromatic Multispectral	0.5 1	13 x 13 13 x 13	1 4	Yes Yes	3 3	– –	(Q4 2004)
Topsat UK	Panchromatic Multispectral	2.5 5	25 x 25 25 x 25	1 4	– –	5 5	– –	(Q1 2004)
CBERS-3/4 China/Brazil	Panchromatic Multispectral	3 5	– –	1 5	Yes Yes	5 5	– –	(Q4 2004)
IRS-P5 India	Panchromatic Multispectral	2.5 10	70 x 70 70 x 70	1 4	Yes Yes	5 5	– –	(Q4 2003)
IRS-P6 India	Panchromatic Multispectral	1 6	70 x 70 70 x 70	1 4	Yes Yes	5 5	– –	(Q4 2004)
IKONOS-3 USA	Panchromatic Multispectral	0.5 2	20 x 20 20 x 20	1 4	Yes Yes	5 5	– –	(Q1 2005)

+ Missions last about 45 days ++ Archive imagery available from previous missions

Table 7–2 Commercial Satellite Imagery (source: courtesy Steve Adam, Canadian Geomatic Solutions Ltd.)

Developed by Raytheon Santa Barbara Remote Sensing in California, this type of scanner is an improvement over Landsat 4 and 5 TM instrumentation. Yet it still provides data continuity with prior Landsat technologies. One improvement in the instrument includes increased spatial resolution of the thermal IR band 6. Other improvements relate to the radiometric calibration equipment and the addition of panchromatic band 8.

Landsat 7, the most recent addition to the Landsat constellation, orbits the Earth at an altitude of about 438 mi (705 km). It has a Sun-synchronous 98° inclination and a descending equatorial crossing time of 10 A.M. According to NASA, acquired images of the world's landmass have been placed into a worldwide reference system catalog. This catalog is comprised of 57,784 scenes, each 115 mi wide x 106 mi long. Each scene consumes about 3.8 gigabits of data (http://landsat7.usgs.gov/outreach.htm).

At the turn of the century, Landsat satellites continued to circumvent the Earth at an altitude of 700 km. They obtain complete coverage of the globe in 233 orbits during a period of 16 days. For the petroleum industry, Landsat images have made great contributions to the field of exploration. Geologists have long been some of the most active users of this technology.

SPOT (France)

Obviously NASA does not have the only satellite-based remote-sensing system. In addition to Landsat, SPOT (Système Probatoire d'Observation de la Terre) provided a significant contribution to the civilian satellite image collection because of its (then) high resolution (10 m). Moreover, the launch of the first SPOT satellites started the debate over concerns of civilian satellites and privacy.[22]

In 1986, France launched SPOT-1 on an Ariane Rocket from French Guiana. This was followed by SPOT-2, SPOT-3, and SPOT-4 in 1988, 1993, and 1998, respectively. SPOT-3 was lost in orbit in 1996; SPOT-5 was launched in May 2002.

SPOT satellites reside in Sun-synchronous orbits that allow them to revisit an exact point above the Earth every 26 days. This allows users to obtain images and data over a given area on a regular basis. This Sun-synchronous orbit takes place at an angle of 22.5° at 10:30 A.M. local time. Thus the Sun's angles on the ground are kept constant during each imaging pass. This means that imagery is acquired at a given latitude with constant illumination.[23]

SPOT satellites use two different sensing systems to image the planet.

Black and white panchromatic images. One SPOT sensing system uses black and white panchromatic images from the visible to the near-IR portion of the EM spectrum. Ground resolution is 10 x 10 m for older SPOT satellites and 2.5 x 2.5 m for SPOT-5. For panchromatic imagery, two digital SPOT sensors are typically used to identify and measure surface features. These include such features as buildings, infrastructure, and roads, which are identified by their physical appearance (size, shape, color, and orientation). Moreover, one set of detectors may be used to measure reflected energy such as visible blue, while another set measures near IR. Two separate detector arrays may even be used to measure energy in two different parts of the same wavelength. (See http://www.spot.com/HOME/NEWS/objective_guide/Part3/Part3.htm for further information.) Once these multiple reflectance values are combined, it is possible to create color images.

MSS sensors capturing green, red, and reflected IR bands. The other SPOT sensing system uses MSS sensors to capture green, red, and reflected IR bands. This system has a ground resolution of 20 x 20 m for older SPOT images and 10 x 10 m for SPOT-5. MSS imagery is acquired by a digital sensor that measures reflectance in many bands. Figure 7–7 (Plate 7–4), provided by Floyd Sabins, was one of the first SPOT images to be produced.

With MSS imagery, remote-sensing satellites measure reflectance in three to seven different bands at once. MSS applications can be used to identify more obscure features, such as chemical properties, mineral content, moisture level, and vegetation types. [An MSS image is minimally composed of the three additive primary colors (red, green, and blue), whose combination can be used to create an image.]

SPOT satellites can also acquire overlapping pairs of images in the same manner as aerial photography. With fixed-position stereo sensors, in particular, these are used to create digital elevation models (DEMs). Satellites like Landsat, however, can only capture overlapping single images from different orbital paths. Commercial SPOT DEMs have a horizontal accuracy of 10 m and a vertical (height) accuracy of 15 m. Depending on the *swath width*, or field of view, SPOT-5 can produce black and white imagery at 2.5- and 5-m resolutions, and color imagery at 10- and 20-m resolutions. DEMs like those produced by SPOT are among the most widely used thematic layers in GIS.

IRS (India)

India has provided excellent contributions to remote sensing through its Indian Remote Sensing (IRS) satellite system. This system provides information in the areas of agriculture, water resources, forestry, ecology, urban sprawl, damage assessment, and geology (http://www.fas.org/spp/guide/india/earth/irs.htm). Of particular interest are the IRS-1C and 1D satellites, launched in December 1995 and September 1997, respectively. Until 1999, these satellites provided the highest commercially available imaging resolution at 5.8 m panchromatic (resampled to 5-m pixel detail). The IRS-1C and 1D satellites are also equipped with:

- Two-band wide-field sensors that cover an area of 774 km^2 in a single image
- Four-band multispectral sensors (red, green, near IR, and SWIR portions of the EM spectrum) that provide 23.5-m resolution multispectral coverage. This is resampled to produce 20-m pixel detail
- A two-channel (0.62–0.68 µm and 0.77–0.86 µm) wide-field sensor that captures images at a resolution of 190 m

Images captured from IRS-2 have contributed to an extensive catalog that covers a broad portion of the globe made available to users all over the world.

The intended launch of IRS-P5, slated for fourth quarter 2003 (see Table 7–2), will be used for terrain modeling applications and cadastral mapping at a scale of 1:5000. Sensors will provide 2.5-m panchromatic resolution and 10-m multispectral resolution with fore-aft stereo capability. IRS-P6, slated for launch late 2004, will have a three-band multispectral camera with a spatial resolution better than 6 m. It will also include an improved version of the four-band MSS found on IRS-1D. This satellite will be dedicated to agricultural purposes.

IKONOS (United States)

As of the writing of this book, the IKONOS satellite produces some of the highest-resolution commercially available images. These are produced from the visible and near-IR portions of the EM spectrum. Launched in September 1999, this satellite provides 1-m panchromatic resolution and 4-m MSS resolution images from blue, green, red, and near-IR bands. Revisit time over the same location is approximately 3 days with an operating swath of 11.3 km. IKONOS orbits the Earth every 98 minutes at a height of 682 km.

Soon after its launch, TransCanada Pipelines used imagery captured from IKONOS to map more than 1500 km of mainline. This was mapped for the purpose of identifying population structures within a buffer of the pipeline's right-of-way.[25]

A pilot test was conducted to determine positional and observational accuracy of IKONOS imagery as compared to digital orthophotography (aerial photos). Steve Adam of Canadian Geomatic Solutions Ltd. found a high degree of positional correlation, falling within 6.5 m of the orthorectified photos. Furthermore, he found that IKONOS imagery missed only five structures along 47 km of pipeline right-of-way (three stretches). Altogether, Adam found a 90% accuracy rate with four minor sources of errors:

1. Shadows in low sun angles. Problem can be minimized by collecting data in late spring or early fall.
2. Structure identification. A common source of error even with orthophotographs. Fortunately, most large commercial and industrial structures can be distinguished from one another.
3. Time lapse between images. The user must remain aware of the time between image acquisition and steady changes in the landscape, infrastructure, and urban sprawl. This is especially important when comparing one vintage source with another.
4. Image rectification. Positional errors on the order of 5 m can occur through inherent positional errors experienced with georeferenced imaging technology.

Among its many uses, IKONOS imagery has been used to provide information during emergency situations. The high-resolution photo shown in Figure 7–8 (Plate 7–5), for example, shows the areal spread of an oil slick. This oil slick took place near the junction of the Detroit and Rouge Rivers in April 2002 when an unknown source of hydrocarbons affected 27 mi of shoreline. Note the dark areas in the center of the image. The high-resolution imagery can be used to observe objects larger than 1 m^2. Like SPOT imagery, IKONOS data can be used for DEM and orthorectified data sets.

It should be noted that there are other systems with very high resolution. DigitalGlobe's Quickbird, an Israeli satellite, also provides imagery at resolutions of less than 1 m. Russia's SPIN-2 satellites provide 2-m panchromatic images. Finally, OrbImages' Orbview-3 is expected to provide 1-m resolution once launched.

Shuttle radar topography mission

The beginning of the end for spirit and trigonometric leveling used in topographic mapping likely took place with the commercial introduction of GPS in the 1990s. If so, the launch of the space shuttle *Endeavor* on February 11, 2000 certainly provides another alternative. This mission, which lasted 12 days, utilized interferometric radar at both C-band and X-band frequencies. It has gathered "the most complete near-global high-resolution database of the Earth's topography."[26]

During the survey, the mission covered 80% of the Earth's land surface to produce about 14,300 cells (1° x 1°) of elevational data.[27] Resolution of the imaging took place at 30-m x 30-m spatial sampling with 16-m absolute vertical height accuracy. It had 10-m relative vertical height accuracy and 20-m absolute horizontal circular accuracy.

The SRTM utilized two radar antennas. One was located in the payload bay (transmitter and receiver). The other was positioned at the end of a 60-m mast that extended from the payload bay (receiver only). With this configuration, two images of the same area taken from different vantage points allow users to produce digital terrain elevation data (similar to DEMs) for both C-band and X-band frequencies.

The SRTM acquired data along 225-km swaths and imaged all of Earth's land surface between 60° N and 56° S latitude. Upon completion of processing the worldwide coverage, the resulting digital topographic map will form a homogeneous data set referenced to the WGS84 global geodetic datum (see chapter 4). Data will be provided in compatible formats for standard GIS software and terrain analysis programs.

Radar sensing

In November 1995, Radarsat-1 was launched by the Canadian Space Agency. As a remote-sensing device, Radarsat is quite different from Landsat and SPOT technologies. It is an active remote-sensing system that transmits and receives microwave radiation. In comparison, Landsat and SPOT sensors passively measure reflected radiation at wavelengths in the visible and near-IR spectrum.

Radarsat's microwave energy penetrates clouds, rain, dust, or haze and produces images regardless of the Sun's illumination, allowing it to image in darkness. Radarsat images have a resolution of 8–100 m. This sensing technology has found important applications in oil slick detection, crop monitoring, and defense surveillance. It has also been used for disaster assessment, geologic resource mapping, sea-ice mapping, and digital elevation modeling.

In July 1991, Europe launched its first radar altimeter (ERS-2), capable of altimeter range measurements of just under 5 m. The ERS-2, however, has been devoted mostly to studies of the ocean, the Mount Etna eruptions in 2001, and other scientific endeavors. With such high-resolution capability, though, there will certainly be uses for this system in the petroleum industry if made available.

Others

Obviously there are a multitude of other remote-sensing satellites that are worthy of discussion, many of which are noted below and in Table 7–2.

Earth Observing System Terra. Earth Observing System Terra (EOS AM-1, NASA) was launched on December 18, 1999 [see Fig. 7–9 (Plate 7–6)]. It contains a wide array of sensors that monitor land cover and use, seasonal and interannual climate variability, natural hazards, and long-term climate changes. Its sensors include:

- Advanced spaceborne thermal emission and reflection radiometer (ASTER)
- Clouds and the Earth's radiant energy system (CERES) (two identical scanners)
- Multiangle imaging spectroradiometer (MISR)
- Moderate resolution imaging spectroradiometer (MODIS)
- Measurements of pollution in the troposphere

OrbView Imagery. OrbView Imagery is obtained from the OrbView-1 satellite and includes severe weather images and global lightning information at 10-km resolution. ORBIMAGE's OrbView-2 satellite provides imagery with 1-km resolution. A future launch of Orbview-3 is expected to provide 1-m panchromatic and 4-m multispectral imagery.

DigitalGlobe's QuickBird. DigitalGlobe's QuickBird was launched on October 18, 2001. Panchromatic resolution is rated at 0.61 m; multispectral resolution is 2.44 m (DEMs).

Imagesat's EROS A1. Imagesat's EROS A1 was launched December 5, 2000 from Siberia. It is capable of producing pictures with a 1-m resolution from its orbit 480 km above the Earth.

The author apologizes for not covering these and others in greater detail, but this would require a more thorough treatise on the subject that extends beyond the scope of this book.

Spatial Resolution

Much of the previous discussion centered on various levels of resolution, all taken from a variety of remote-sensing images. But when one talks about 1-m, 10-m, and 1-km resolution, what does it all really mean?

Spatial resolution refers to the size of the smallest object or ground feature that can be distinguished in an image. Ground resolution, as such, concerns the smallest area on the Earth's surface from which reflectance measurements can be made. (This assumes measurement by a given sensor at a given altitude having a specified swath width and scan time.)[29]

One-meter imagery like that acquired from the IKONOS, Quickbird, and EROS-1 can be used to identify and map objects. These may be the size of automobiles, utility equipment, trees, roads, petroleum infrastructure, and outcrops. (Typically, ground unit measurements are taken in meters.) Imagery with a resolution of 10 m, like that provided by SPOT, can be used to locate and map prominent outcrops, property boundaries, buildings, roads, and facilities. Imagery ranging from 20–30 m, obtained from Landsat satellites, can be used to locate forests, fields, city centers, suburbs, and large factories.

Imagery with 80-m resolution can be used to locate regional geologic structures or access vegetative health over a large area. Finally, 1-km imagery can be used to access vegetation across regional areas. It can also be used to track small-scale events like land damage induced through storms or insect infestation.

Before choosing imagery for GIS, one must determine both the sizes of individual objects and the areal distribution of all the regional features to be analyzed. Only then can the most appropriate mixture of imagery with appropriate resolution be chosen.

Obviously there are two considerations here: cost and quality. High-resolution imagery is more expensive per unit area than low-resolution imagery. On the other hand, low-resolution imagery may miss features that you are trying to locate, in turn leading to poor decision making. Consider a case in which low-resolution imagery is used to determine the route for a pipeline corridor, but fails to recognize impassable sections like swamps and bogs. The cost of route reengineering once the work has begun will outrun the cost of high-resolution imagery.

In many cases, companies will simply send survey personnel out in the field to directly map the route beforehand. However, the costs and lost time involved with this work can also be unfavorable when compared to the cost of high-resolution imagery.

Spatial resolution also impacts the efficiencies and costs of computer processing and storage. For example, a raw 10-m resolution SPOT panchromatic image of an area 60 x 60 km in size can result in a 36-megabyte file. A 1-m resolution panchromatic image of the same area may require 3–4 gigabytes (SPOT). (As a rule of thumb, a 50-m resolution image can result in 4 times the amount of data as a 100-m resolution image.[30]) Advances in computing power, storage capacity, and raster cell compression techniques, however, do not make this as much a concern as it once did.

Keep in mind that most EM-based remote-sensing sensors capture images in raster format. Thus each pixel represents not only the wavelength values (attributes) but also the ground sampling size (spatial resolution). As such, determining the best spatial resolution involves several design tradeoffs. These involve the accuracy of not only *seeing* the feature but ensuring it is within an acceptable distance of the stated coordinate. If the feature is smaller than the pixel size, for example, it will not be available for analysis in the GIS. Additionally, if a 1-pixel cell represents a 20-m^2 area, then it is possible that an object in the image can be misplaced by 20 m in any direction. Such displacement can happen through side shifting, which is a function of pixel topography.

With this said, users must use good judgment to mix and match high- and low-resolution imagery to fit their needs.

Image Attributes

An *image* is a binary file of unsigned integers that does not contain floating point or negative values.[31] The flexibility of an image is its ability to hold one or many bands, with each band containing different information in a raster format. A basic example is an image that holds several different spectral ranges of light. One band contains the red component of light, the other green, and the last blue. Thus, every pixel in this image holds three pieces of data for the same location. Both single-band and multiple-band images are stored in one file. The header and color-map information are sometimes stored in separate files.

There are essentially two types of images: one-band and multiple-band. A one-band image may represent black and white, grayscale, or color data. In black and white images, for example, each pixel uses one bit of information. The value of the background is usually 0 and the foreground is usually 1.

One-band color images are called pseudocolor images. They consume 4 or 8 bits of data per pixel, including the color map. The color map contains an index number along with red, green, and blue values that make up the color for that index. Pseudocolor images are limited to a total of 256 colors out of the total spectrum (in the case of 8 bits). The color map may be contained within the image file, or it may be in a separate file.

Multiple-band images may be displayed as grayscale or color. Each band in this type of image usually contains 8 bits of data per pixel although some grayscale images have 32 bits per pixel. It is possible to display one band as a grayscale image or combine any three bands into a color image. When three bands are displayed, a true-color image is composed. As such, the three bands can be used to give a value of red, green, and blue for each pixel. The color resulting from that combination will be the color of the pixel. These are called true color images because they can show up to 2^{24} or more than 16 million different colors.[32] Obviously, the human eye cannot distinguish all of these shades from one another.

Geographic Registration

To ensure a seamless overlay in GIS, all data layers must be georeferenced to the same coordinate system (see chapter 4). This is true whether they are vector or raster (remote-sensing) data. Vector data are precisely defined to some Cartesian coordinate system. But raster data and images, including Tiffs, Gifs, Jpegs, and Mrsids, can only be positioned in a view based on the attributes of the image pixels. Each cell is defined by a row and column number (see chapter 3, Arcview 3.3 Help File).

These image formats may store georeferencing coordinates in the header of the image file itself (Erdas IMAGINE files, MrSID images, GeoTIFFs, and so forth). Or they may use a cross-link to an additional header file (Hdr, Bil, Bsq, and Bip). The header and data (image) files can usually be identified with one another by comparing the extension. For example, the header file for the image file mytown.tif would be called mytown.tfw (Arcview 3.3 Help File).

Many GIS programs store coverages in real-world coordinates. To display images with coverages or shapefiles, then, it is necessary to establish an image-to-geodetic transformation. This converts the image coordinates to real-world coordinates. This transformation information must be stored with the image.

Another topic that GIS users must consider includes the means by which remote-sensing images are geographically matched with other themes. As such, registration refers to aligning two data sets. This can be alignment of one image to another image, or an image to some precisely defined set of GPS benchmarks.[33] For GIS, all data sets must be geographically matched so that accurate spatial analysis can be conducted. Problems associated with roll, pitch, and yaw from either an airplane or satellite must be accounted for.[34]

There are basically three forms of registration: image-to-layer, image-to-image, and GPS-to-image. The first type matches the image to the map. This is accomplished by finding points or features that are clearly identified on both the image and the map layer (perhaps within a GIS). Such features might include roads, pipelines, or natural landforms. The second type matches one image with another through reference points obtained from an already geocorrected image. The third type references exactly defined ground points using a GPS receiver to the image that has clearly recognizable counterparts.

Always remember that raw data do not represent locations on the surface of the Earth unless the data carry some reference to the ground location.

Difficulties with Data

Importing remote-sensing data into a GIS requires a lot of patience, a strong knowledge of data formats, aptitude with conversion programs, and the right software. In fact, retrieving remote-sensing data is perhaps one of the most difficult activities in GIS.

There are many sources of data available for download, both free and commercial. Unfortunately, many do not provide clear-cut instruction on how to download, import, and format the images. In fact, industry standards that should have been put into place years ago have yet to be realized.

Moving towards this goal, the Spatial Data Transfer Standard (SDTS) was developed to address the transfer of geospatial data among computer systems. But there are still drawbacks. "The Raster Profile of SDTS, for example, can only be used to incorporate remote-sensing data from a proposed swath standard. Yet the SDTS Raster Profile is a transfer standard, while the proposed swath standard is a content standard."[35] Based on this premise, a SDTS raster profile can be used to transfer remote-sensing swath data. However, there is no overlap between the standards because both have different aspects of data standardization.

If this has not confused the reader, the following procedure in Box 7–1 for transferring and converting SDTS files to ArcView shapefiles certainly will.

Box 7–1

Converting SDTS Files to Arcview Shapefiles

Before starting any work, it is easier to download all the files used in the conversion steps. They come from a variety of sites; Internet address, file names, and links can be found below. A standard browser can be used to download files from any of the sites, including the FTP sites.... You need the following files from this site:

1- 00MasterDD_LRG_SDTS_tar.zip (master data dictionary for 1:24,000 data)
2- Desired data file(s); these are organized alphabetically, by quad name, with the following conventions:
SDTS transfers are stored as compressed "tar" archive files. Each transfer contains a complete set of files, for a specific layer and version, covering a quadrangle. The transfer naming convention is as follows:
XXX.layer.V.sdts.tar.gz.
- XXX = unique quadrangle identifier which consists of quad SE coordinates preceded by a D
- layer = data category name abbreviation (such as bd for boundary) followed by 0 (zero) and "s" (for 7.5 minute) or "f" (for 15 minute)
- V = file version (1-current, 2-historic)
- SDTS = denotes that the data is in SDTS format
- tar = represents "tar" file
- gz = refers to the "gzip" compression

.... There are a number of ways to extract or uncompress the files. WinZip used with the "TAR file Smart CR/LF conversion" turned off will work. Another option is to use gunzip with the -d option, then tar with the -vxf option to decompress and extract the files....

After uncompressing the files there will be many files in the directory. These files are discussed in detail in the complete SDTS Transfer Description....The next step is to convert the DDF files to DLG format, using the SDTS2DLG.exe and other BLM programs. This program and all the BLM executables should be run at the DOS prompt, preferably in the same directory as the .DDF files being converting. ***Please*** *pay attention to the notes below on naming conventions, as they will help you avoid problems when running the ArcView scripts later.*

1- Type SDTS2DLG and enter the required info when prompted:
a. First enter the first 4 characters (TR01 for transportation files, HY01 for hydrology, etc.)
b. Enter an output file name (hyav, rdav, etc)
c. Enter the digits in positions 7 and 8 of the file being converting. This is usually 01 but may vary for 1;100,000 scale or other multiple layers. The file names will contain LE, NA, NE, NO, NP, or PC, depending on the type of geographic feature being referenced such as line (LE) or polygon (PC). See table 2 in the SDTS Transfer Overview for more information on these files.
d. This program creates two files. If the output file name was hyav, the files will be called hyav.dlg and hyav.dla. Whatever names you choose in this program, you need to consistently use the same naming conventions for steps 2, 3, and 4 below as well or the Avenue scripts will not run properly.

Source: U.S. Fish and Wildlife website, (http://www.fws.gov/data/gisconv/sdts2av.html)

Importing DEMs, orthophotographs, and other formats also have various levels of difficulties, requiring time and experience before images can be properly integrated into a GIS project. Shuttle radar topography data, for example, can only be downloaded via file transfer protocol (ftp). Once downloaded, data formats can obviously only be used with GIS software (of course), but certain nuances must be dealt with to use the data. For example, ArcView can display the DEM data directly only after the file extension *.hgt has been renamed to *.bil. And if a user needs access to the actual elevation values, the DEM data must be converted into an ARC/INFO grid (USGS).

Fortunately, there are efforts underway to improve the integration of remote sensing and GIS technologies. The OGC, for example, has been working with software vendors, earth imaging vendors, and database software vendors. They are striving to reach agreement on technical details of open interfaces that will allow these systems to work together, over the web. Specific to remote sensing, the Open GIS Grid Coverage Implementation Specification addresses *gridded* data. These gridded data differ from the definition used by some GIS software vendors and can include satellite images, aerial photos, and digital elevation data.

Although a great deal of time must be spent locating, downloading, importing, and formatting remote-sensing data, the benefits far outweigh the costs. As such, it is recommended to involve a remote-sensing/ GIS specialist early on in any project that requires the use of raster imagery.

Examples

The final topic concerns applications in the oil industry. In many petroleum activities, especially in the upstream sector, remote sensing is used in tandem with other technologies to gain a better understanding of a particular problem. Thus, its use requires a multidisciplinary approach.

Take one of the most common remote-sensing applications in our business, exploration. Dr. F.F. Sabins, a former researcher for Chevron and an adjunct professor at the University of California–Los Angeles, outlines a typical exploration program as follows:

- Regional remote-sensing reconnaissance with small-scale land mosaics covering large areas or even entire sedimentary basins
- Reconnaissance geophysical surveys including magnetics and gravity
- Detailed remote-sensing interpretation designed to map geologic structures and identify prospect locations
- Seismic surveys needed to look at the subsurface
- Wildcat drilling[36]

Sabins credits remote sensing for the successful exploration of Saudi Aramco's Central Arabian Arch.[37] Although "Saudi Arabia has the world's largest oil reserves....it is surprising to realize that the kingdom was relatively unexplored for oil by the late 1980s."

To aid the kingdom in finding new oil deposits, Saudi Aramco then requested assistance from the remote sensing research group of Chevron. Major objectives of this 1988 study included regional geologic mapping at a scale of 1:250,000, which utilized seven Landsat TM 2-4-7 images. It also involved the interpretation of local anomalies that may have evolved through expressions of subsurface structures. This work was conducted under the supervision of L.E. Wender of Saudi Aramco and Sabins.

Once the study was completed, Saudi Aramco used the Landsat interpretation in combination with seismic studies to make a large discovery. It initially flowed 2984 bbl of oil, 1180 bbl of condensate, and 26.2 mcf of gas per day.[38]

Remote sensing also plays a variety of roles that range from oil spill risk assessment to least-cost path analysis for pipeline routes. As such, there are five primary functions of remote-sensing images for GIS-based oilfield work. These include base-map backdrops, map-vector updates, geologic mapping, change detection, and 3-D modeling. Base-map backdrops are perhaps the most primitive application, where geophysicists and others use orthophotographs as aides in analyzing the terrain around seismic shot point profiles and drilling locations.

Drafting departments also find great use for remote-sensing images by overlaying scanned or digitized maps on top of satellite and aerial photos. This is done to find inaccuracies or update maps. New roads may become clearly apparent in an image, for example, allowing the drafting specialist to manually update line vectors. Thus updates are accomplished without sending a crew to the field with GPS receivers.

Change detection has become increasingly useful as well, especially in the petroleum industry. Satellite imagery may be acquired on a periodical basis, allowing users to analyze changes in land use, geomorphology, and postdisaster effects. Once these images have been rectified, the GIS can compare the values of corresponding pixels, denoting whether some form of change has taken place.

Finally, with 3-D modeling, a satellite image can be draped over a DEM to create a 3-D view of the Earth. Geologists have found great utility with this tool, as it allows them to move beyond 2-D maps to make 3-D analyses of Earth structures.

Oil spill decision support

To reduce the response time and qualify the decision-making process for contingency planning, remote-sensing and GIS technologies can be integrated and utilized. They prove useful for detecting, mapping, evaluating, and modeling oil spills in marine environments. One group of researchers, funded by Malaysia's Ministry of Science,

Technology, and Environment, was headed by Dr. Shattri Mansor from the University of Putra Malaysia, Malaysia. This group developed an early warning system to identify vulnerable coastal locations once a spill has taken place.[39]

The system, known as Oil SCAN, provides a useful tool to identify, delineate, and estimate selected response technologies used in marine oil spill clean up. This system takes into account given environmental conditions, changes in oil characteristics, and operational considerations in the affected area.

The Oil SCAN system consists of two main components, each having a specific use for decision makers (see Fig. 7–10):

1. A detection and classification tool (DCT). The DCT provides data preprocessing for geometric and radiometric correction. It also provides postprocessing, such as image filtering and texture analysis for oil spills.
2. A decision support tool (DST). The DST allows users to incorporate ancillary data. Such data include remote-sensing images, scanned oceanographic charts, coastline vectors, ESI maps, and response team information for spatial analysis and modeling. This tool provides a systematic and seamless data management capability that incorporates trajectory modeling for risk assessment and contingency planning. DST outputs consist of custom-formatted images with vector and object-based overlays. The analysis is presented based on identified map features built on processes specific to the application.

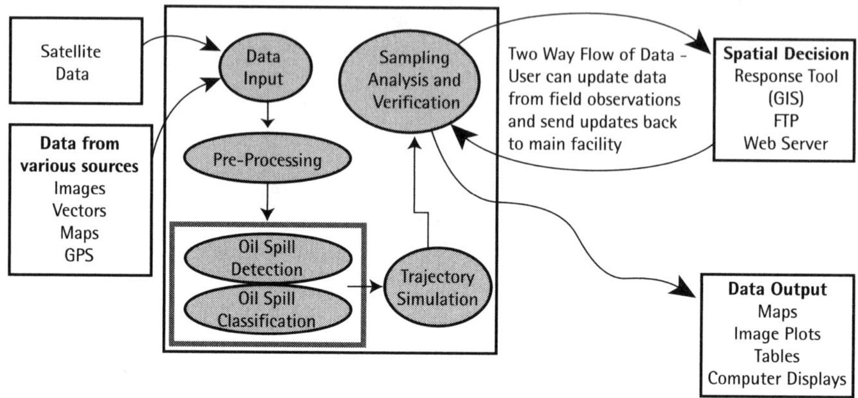

Fig. 7–10 Oil Spill Decision Support Model
(source: Dr. Shattri Mansor, University of Putra Malaysia, Malaysia)

Radar images and optical satellite data can be used to detect oil spills. Figure 7–11a, for example, shows a RadarSat SAR image acquired over the Straits of Malacca on October 26, 1997. In this image, dark areas define both the spill area and spill look-alike areas, requiring further processing to eliminate the latter. Figure 7–11b (Plate 7–7) shows a filtered and classified radar image. The spill look-alikes were eliminated. The black area is the spill area, the green area is a high-pollution area, and the yellow area is a low-pollution area.

Radar image processing can be divided into two steps: preprocessing and postprocessing. Preprocessing involves antenna pattern (APC), radiometric, and geometric corrections. Postprocessing serves to delineate the oil spill through several steps. These include image enhancement, texture analysis, dark-slick detection, feature extraction, scaling, and filtering. First, the program calibrates a radar brightness image generated from the SAR image. Second, the program uses gain offset and scaling to generate a radar brightness channel from the input scaled radar channel. Third, the software builds a texture set for all pixels on the image.

At this point, the image can be used to classify the oil spill using a supervised maximum likelihood algorithm [see Fig. 7–11b (Plate 7–7)]. Speckles appearing on SAR images are a natural phenomenon generated by the coherent processing of radar echoes. The presence of speckles not only reduces the interpreter's ability to resolve fine detail, but also makes automatic segmentation of such images difficult. The gamma map filter is primarily used on radar data to remove high-frequency noise. Parameters for the spill detection modules were set in advance, and the model was initiated.

For oil spill risk assessment, it is important to locate the extents and predicted movements of an oil slick so that a proper plan can be formulated to deal with the problem. *The model therefore consists of locating and monitoring an event through remote sensing, then placing the imagery into a geographical context so that GIS can assess the situation.*

To evaluate the model, the University of Putra Malaysia researchers analyzed historical SAR data from prior spills and integrated this information into the system. This allowed investigators to determine the dynamics of prior slick movements. In turn, they derived a drift prediction model based on the following equations.

$$r(t) = r(t_0) + \int_{t_0}^{t} \overline{V}_{oil}(t')dt' \qquad (7.1)$$

where

$r(t)$	is the vector coordinate of a slick at time t after its initial spillage at time t_0
$r(t_0)$	is the initial spillage situation at time t_0
V_{oil}	is the spill drift velocity
$(t')dt$	is the time duration of the oil spill

Combining both the current and wind effects and the relationship of the movement of the center of mass of the water, these factors can be expressed as:

$$V_{water} = dr/dt = Uc + 0.035\ Uw \tag{7.2}$$

where

V_{water} is the velocity of center of mass of the water
Uc is the current speed near the surface
Uw is the velocity of wind at the surface

Regardless of the physical properties of oil and water, the size of the spill and its spreading tendency, the resultant vector may be estimated as:[40, 41]

$$\overline{V}_{oil} = (0.035\ Uw)\ \overline{Kw} + (0.56\ Uc)\ \overline{Kc} \tag{7.3}$$

where

$\overline{Kw}$ is the unit vector in the direction of wind drift
$\overline{Kc}$ is the unit vector in the direction of current
V_{oil} is the spill drift velocity
Uw is the wind speed at sea surface
Uc is the current speed near the surface

Then deriving a formula for wind factor, *WF*, it is found that:

$$WF = Uw / V_{10} \tag{7.4}$$

Most meteorological observations for wind speed are taken at 10 m above the sea surface. The velocity of wind at the surface (Uw) and the wind velocity at 10 m (V_{10}) are related by a wind factor (WF). The wind factor can be evaluated by the following formula, where Φ is the latitude of the spill.

$$WF = 0.04/(V_{10} /\sin\Phi)^{1/2} \tag{7.5}$$

Thus, wind, wave, and currents form the main parameters of this model. All are utilized in the GIS program to assess the movement of local oil spills.

Least-cost pipeline analysis

Another excellent example of utilizing remote sensing in an almost pure GIS context involves a least-cost path network analysis. This analysis was conducted by Bechtel and NASA for a pipeline segment running from the Tengiz oil field, Kazakhstan to Novorossiysk on the Black Sea (Russia). Described by Morain and Baros, this $2.65 billion, 900-mi long pipeline came on-line in November 2001.[42] It is expected to handle 1.5 million BOPD.[43]

The route skirts the north end of the Caspian Sea, then runs through Astrakhanskaya, Komsomolskaya, and Kropotkin to the Novorossiysk area. Part of the pipeline, from Tengiz to Komsomolskaya, was in operation before construction began on the new segment.

The remaining segment, however, involved several pipeline stretches that ranged from predominantly low relief to semimountainous areas.

According to authors Sandra Feldman and Ramona Pelletier of Bechtel and NASA, a 50-km pilot project located near the northern end of the Greater Caucasus Mountains was chosen for analysis. The goal was to determine if remote-sensing imagery, in combination with vector-based data, could be used to determine the cheapest path possible. "In theory, the cheapest route for building a pipeline is a straight line between two points."[44] However, obvious costs associated with topography, surface geology, river and wetland crossings, road and railway junctions, and urban districts precluded this assumption.

For the study, 25-m resolution Landsat TM imagery, 1:500,000 scale topographic maps, and 1:500,000 scale geologic maps were all integrated in the GIS. Technicians digitized road and drainage features with ARC/INFO. Topographic contours were converted into DEMs.[45] All vector layers were then converted into raster format for spatial integration.

To simplify interpretation for pipeline personnel, each geologic unit was classified as consolidated hard rock or unconsolidated. Land use information, slope map, and other geographic features digitized earlier were then converted to raster format. This was accomplished using a GIS program developed for the U.S. Army Corps of Engineers. As a result, individual themes based on a raster format were standardized.

Based on modified algorithms commonly used in GIS for drainage basin analysis, nodes and endpoints were emplaced to constrain the path. "The procedure involved resampling each of the raster layers into 25 meter cells." For every layer, each cell was assigned cost values based on predetermined construction costs associated with the type of passage. For example, hardrock or hilly areas were assigned higher values than unconsolidated soil or flat terrain. Feldman and Pelletier also assigned a cost value of 10 times the baseline construction value for urban and industrial lands.[46]

The least-cost algorithm then added the cost values of each cell to compute cumulative costs and develop a cumulative cost surface map over the entire pilot section. Represented much like a topographic map, peaks represented the most costly areas, and valleys the least.

The authors then conducted two analyses: one using manual calculations and the other raster-based GIS. For the former, they drew a straight line between each of the points in the pilot area to determine the cumulative costs for the shortest route. Total cost was then calculated by adding these values together. For the latter, GIS specialists performed a network analysis across the nodes. This resulted in a more economical route based on looking at individual cell values rather than looking at each stretch. Moreover, the manual straight-line method crossed more urban areas, roads, and hardrock areas than the GIS model did. Although the distance was shorter for the straight-line method at 42 km, the GIS analysis showed it would be cheaper to build a 51-km least-cost route.

Conclusion

Without question, remote-sensing imagery augments the usefulness of GIS. Information derived from this source is global, easy to acquire, and often cheaper than alternatives. Furthermore, the data provide views that extend our senses, allowing users to more closely approximate true reality. Finally, the ability to fly over the same location repeatedly, or in most cases investigate a location within a reasonable period of time, makes it an invaluable tool for tracking change. Although the industry needs to find a means to make it more user-friendly, the benefits clearly outweigh the costs.

References

[1] Sabins, F.F. 2000. *Remote Sensing: Principles and Interpretation.* 3rd ed. Freeman and Co. p. 1.

[2] Foresman, T.W. ed. 1998. *The History of Geographic Information Systems: Perspectives from the Pioneers.* Upper Saddle River, NJ: Prentice-Hall. pp. 21 and 148.

[3] Sabins, F.F. (Plate 12C. Used with permission).

[4] NASA website (http://rst.gsfc.nasa.gov)

[5] Morain, S. and S.L. Barrow. eds. 1996. *Raster Imagery in Geographic Information Systems.* Santa Fe: OnWord. p. 8.

[6] Sabins, p. 3.

[7] Ibid. p. 5.

[8] Morain, p. 9.

[9] Sabins, p. 4.

[10] Morain, p. 9.

[11] Ibid. pp. 8–11.

[12] Sabins, p. 3.

[13] Ibid.

[14] Pidwirny, M.J. Dept. of Geography, On-line, Okanagan University College.

[15] Prost, G.L. 1994. *Remote Sensing for Geologists: A Guide to Image Interpretation.* Case postale, Switzerland: Gordon and Breach Science Publishers. p. 23.

[16] Sabins, p. 40.

[17] Ibid.

[18] Grendorffer, G. and R. Bill. 1994. *Digital Orthophotos for Mapping and Interpretation in Hybrid GIS Environments.* Stuttgart University. EGIS.

[19] USGS.

[20] NASA.

[21] Sabins. (Plate 6. Used with permission).

[22] Adam, S., 2002. Canadian Geomatic Solutions Ltd. Personal communications.

[23] *SPOT User's Handbook.* Vol. 1. 1988. Reston, VA: SPOT Image Corp. (http://www.spot.com/home/news/publi/publi.htm).

[24] Sabins. (Plate 8. Used with permission).

[25] GITA 2000 Proceedings, p. 7.

[26] Jet Propulsion Laboratory, NASA. (http://www.earthsensing.com/rssg/systems.html).

27 Federation of American Scientists. (http://www.earthsensing.com/rssg/systems.html).

28 Bordi, F., S. Neeck, and C. Scolese, Computer Sciences Corp., and NASA/Goddard Space Flight Center, 48th Congress of the International Astronautics Federation, Torino, Italy (October 6–10, 1997).

29 Morain, p. 19.

30 Foresman, p. 167.

31 "How To: Understand General Information about Images." 1993. *ESRI Support Online*, White Paper, Article 11633. (June 11).

32 Ibid.

33 Morain, p. 34.

34 Ibid. p. 26.

35 Proposal for a National Spatial Data Infrastructure Standards Project. 1997. Remote Sensing Swath Data Content Standard. NASA (September 23).

36 Sabins, p. 339.

37 Sabins, 1997.

38 Sabins, p. 355.

39 Mansor, S.B. et al. 2002. "Oil Spill Detection and Monitoring from Satellite Image." Proceedings of Map Asia. Bangkok, Thailand (August 7–9).

40 Mansor, S.B. et al. 1999. "Remote Sensing and GIS Solution to Oil Spill Emergency Response." *Journal of the Malaysian Surveyor* 34:4. pp. 10–12.

41 Assilzadeh, H. et al. 2001. "Early Warning System for Oil Spills in the Straits of Malacca." 2nd International Conference on the Straits of Malacca. (October 15–18).

42 Morain.

43 *Oil & Gas Journal Online.*

44 Morain, p. 126.

45 Ibid. p. 129.

46 Ibid. p. 130.

8

The Art of Presentation

A good map tells a multitude of white lies; it suppresses truth to help the user see what needs to be seen.

—Mark Monmonier

GIS's main role is to support smart decision making through spatial analysis. Yet if the analyst's line of reasoning cannot be clearly expressed, then all of the time spent collecting, preparing, and analyzing the data will come to nothing.

There are well-established, time-honored guidelines for delivering concise and well-designed communiqués. In developing a good GIS presentation, it is important to think about the audience first. Determine what information should be provided second. Organize the GIS project third. Then apply the proper style last.

With some education, these guidelines can be quickly applied to static media such as paper maps or adapted to desktop mapping and even web-based GIS. Moreover, the ability to directly interact with GIS, aided by easy manipulation of categories, classifications, and style, provides new opportunities for the analyst and the audience to communicate in real time. Add the ability to work with 3-D renderings and time-lapse animations, and the GIS user certainly has a powerful toolkit with which to present interpretations and recommendations.

Background

As recently as 10 years ago, the art of presentation fell under the department of drafting. In cooperation with the manager, engineer, or geologist, this department produced graphical outputs and reports under controlled conditions. Obviously, the drafting department had to maintain and support knowledgeable individuals who understood cartographic principles of projections and generalization. Drafting personnel therefore could guide the petroleum professional towards delivering the best output, with preference given to content, geographic correctness, and visual appeal.

To uphold standards and consistency in publication quality, many oil companies went so far as to enforce well-defined style guidelines. These guidelines were used for the graphical representation of assets and objects. For instance green was used to represent oilfields, or dot patterns represented sandstone.

Several improvements in technology have changed this scenario. The dissemination of computers in the corporate office has been accompanied by advances in off-the-shelf mapping and graphical software. The expansion of e-mail and the Internet have also changed the way in which presentations can be distributed. Almost everyone can produce a map that can be put forth for public consumption.

Unfortunately, a little bit of knowledge can be a dangerous thing. It is true that new technology has given users the ability to build beautiful presentations. Without a requisite understanding of cartographic principles and proper map design (style), however, presentations can lead to poor decision making. Always remember that materials derived from GIS data will misrepresent reality if not intelligently constructed.

Think back to the discussion on 2-D map projections in chapter 5. When one tries to project data derived from an Earth-centered, Earth-fixed ellipsoid onto a flat piece of paper, problems arise. Sacrifices will have to be made to illustrate the main objective, whether it involves distance, area, direction, shape, or some combination thereof.

Assume a WGS84 geodetic computation is used to determine the linear measurement along some feature. For example, the running length of the U.S. Gulf Coast tanker route along the shoreline of Texas and Louisiana is measured. However, when the information is delivered in paper form to the audience in a Mercator map projection, complications occur, because the map will not simply support the calculation. Mercator maps are designed for navigation, not distance measurement.

Suppose someone then used this map as a quick means to approximate the length of a pipeline from the East Banks Block in the Gulf of Mexico to Port Arthur, Texas. This person would unknowingly add about 25 mi onto the actual length (a 14% error). This in turn could introduce a significant mistake if used for economic calculations (see Fig. 8–1). The upper map in Figure 8–1 shows the distance based on the WGS84 ellipsoid, while the lower map shows the Mercator projection.

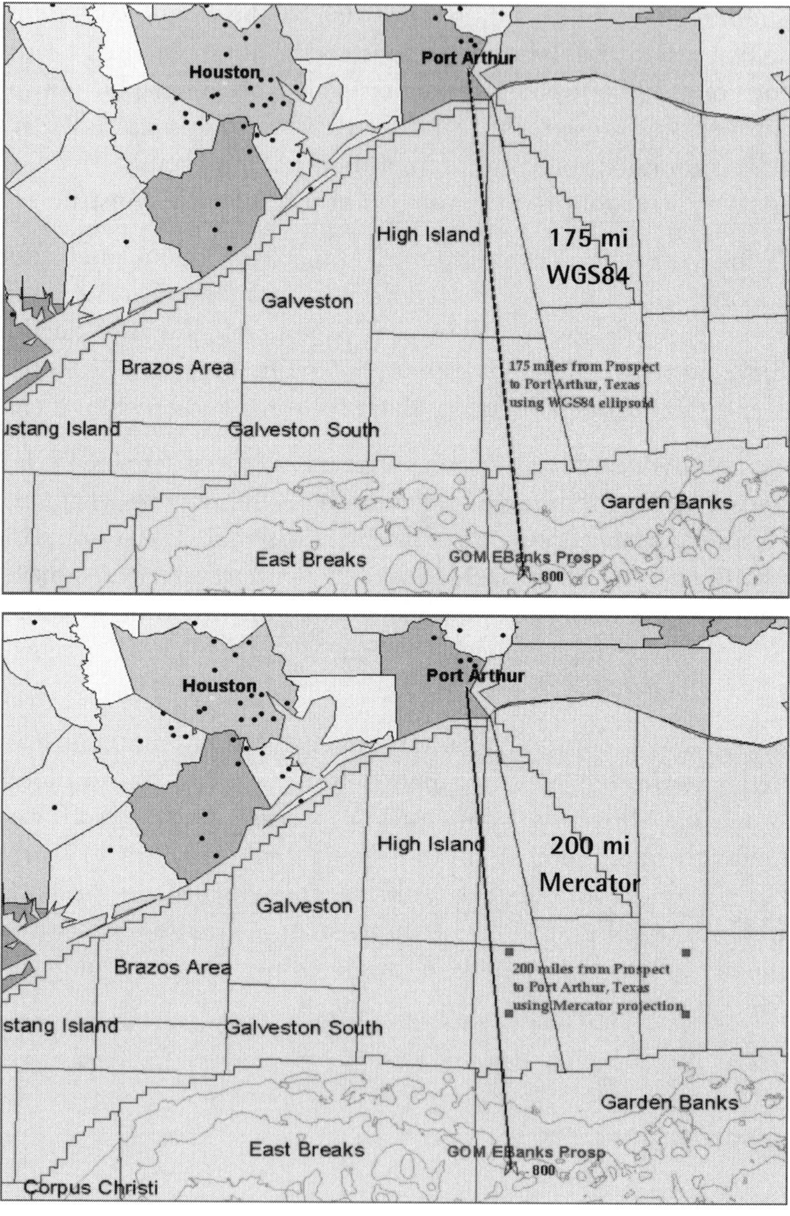

Fig. 8–1 Differences in Distance Using the WGS84 Ellipsoid and Mercator Measurements

Although the inherent accuracy of the calculations would remain correct in the GIS, on a Mercator map projection, this information would be lost to the manager. The manager would assume the map was provided in an accurate format. This is another good reason why geodetic education should be made compulsory for petroleum professionals.

Another example can be related to the statistical representation of information represented in map form. Referring back to Figure 3–5 (Plate 3–2), one can see clearly how different statistical classifications can lead to various interpretations that most certainly would affect marketing activities. The ease in which GIS can take a theme and subdivide data into classes, for example, can lead to huge errors if one does not understand spatial statistics.

There also are problems associated with the selective exclusion of otherwise appropriate information. Leaving off pertinent data can occur through the deletion of point, line, or area entities. Such erroneous results can seduce management into investing in an exploration project that more risk-averse companies would avoid.

The Railroad Valley play in Nevada, for example, has some extremely prolific wells.[1] One could, however, present management two different maps of this area and have different responses. Suppose the first map had *only* producing oil wells denoted. It might receive a different response than the second map (see Fig. 8–2), which showed all producers, abandoned wells, and so forth. Investment in this area might then be reconsidered, as management would have a clearer indication of drilling risk.

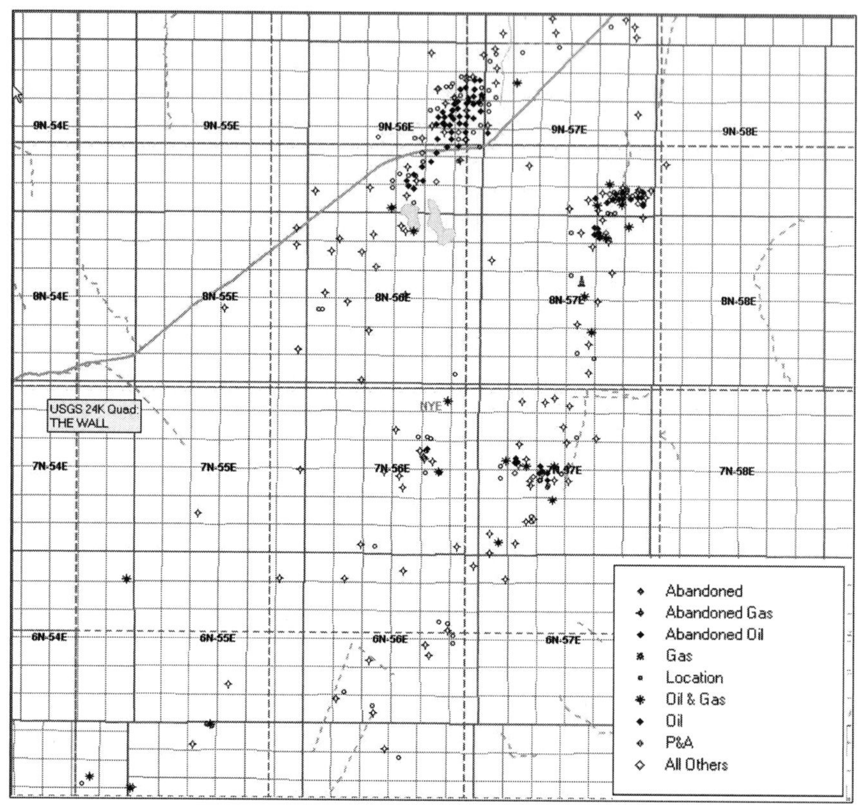

Fig. 8–2 Producing, Abandoned, and Other Wells (source: PennEnergy Data2; www.pennenergydata.com)

This aside, when the analyst applies educated decisions to the final output, computerized cartography can provide a truly visual and communicative means of presenting the interpreter's thoughts. With GIS this often occurs almost upon completion of the analysis, because many outputs are direct expressions of the spatial study that approximate real-world models. By tweaking map views through data reclassification or a simple change in the map projection, the audience's attention can then be directed towards the main focus of the analysis.

For example, the map in Figure 2–7 (Plate 2–5) is based on the *Oil & Gas Journal's* Nelson Complexity Index.[3] It contains no text or distracting features other than data points, which are distinguished by color and size, along with political boundaries and a legend. Yet its apparent simplicity is actually misleading. First, by qualifying the relative cost of each refinery through classes based on standard deviation, the reader is led to believe that an interval system is used to differentiate classes rather than underlying ratios (see chapter 3). Then by applying a Hammer-Aitoff projection, the map view directs the audience's attention from west to east. This presentation more clearly portrays the trend of complex refineries in Western Europe as compared to the less capital-intensive refineries in Russia.

Keeping these issues in mind, it is the map preparer who must decide what data should be included and excluded. The person who prepares the map also determines which projection should be used and what formatting styles should be applied. GIS is useful due to its inherent interface with computerized cartography. As such, it provides an unprecedented opportunity for supporting and presenting an analysis through its ability to map and remap geographic information. It can redefine styles with extreme speed and flexibility.

To provide useful guidelines on the subject of presentations, this chapter will discuss a paradigm shift that is taking place in GIS applications. It will then turn to a discussion on the integrative role of visualization and communication. Next, straightforward guidelines will be provided on how to develop an effective presentation. Finally, examples of 3-D, time-lapse, and Internet-based mapping applications will be described.

Further recommended readings include Mark Monmonier's *How to Lie with Maps*, Christopher Jones's *Geographical Information Systems and Computer Cartography*, Mark Zeiler's *Modeling Our World*, and David Greenwood's *Mapping*.

A Presentation Paradigm

Almost all historic discussions on cartography relate to some form of static presentation, usually in paper form, with supporting tables, graphs, and charts. In this old-world paradigm, all nonessential items and a great deal of supporting data have to be discarded. This is due to limited paper size, time constraints, technical barriers, and lack of human resources. *In a GIS, however, there is 'virtually' no competition for space.*

Because a paper map contains limited space, portraying information is actually a competitive affair that forces the cartographer to eliminate a great deal of useful information.

Instead the paradigm shift moves towards the organization of multiple map views with numerous themes called upon from a database. In other words, limitations now depend on the user's ability to collect and properly format data into successive views supported by themes that can be turned off and on.

Improvements include the ability to produce, intermix, and transfer a variety of layers. Other advances allow the user to flexibly change symbol types and text placement and build mosaics of charts, tables, and graphs. Thus the user can dynamically produce, redraw, or statically fix geographical information.

An interactive GIS presentation might be given to an engineering group, for example, aided by a laptop, projector, and a basic knowledge of cartography. This type of presentation will allow one to react to any question with a custom answer quickly and clearly. Moreover, with some preparation, temporal changes such as 4-D animations can be shown frame by frame, much like a successive series of snapshots. These animations might depict the dispersion of hydrocarbons from an oil spill, for example, or even changing refining patterns across the United States.

Now add Internet-based GIS, and the information becomes the domain of all users across the corporation, with predefined presentation elements put into place by the GIS designer. With these kinds of utilities, never has it been more important to apply cartographic principles.

From the virtual to nonvirtual

Every GIS analyst or presenter has a problem. How does one move from the virtual world of visual thinking in a spatial context to the real world of visual communication for corporate or public education? To answer this question, we must come to terms with the way *visual thinking* and *visual communication* interact with one another.

Spatial visual thinking can be thought of as the analysis of 3-D space, often in the context of time, utilizing one's interpretation of data and information. Visualization therefore comes from the human ability to identify patterns and structure information. By this definition, the direct process of visual thinking takes place solely in the mind, aided by various tools of visual communication.

Take the extreme case of visual thinking used in geologic analyses. Most geologists do not produce their most definitive models on a piece of paper or a computerized mapping program. Instead, they are taught to visualize depositional regimes and geologic structures as images and concepts in the mind. They only utilize computer programs as decision-support tools and vehicles for communication.

It is the geologist, for example, who must first decide whether the prospect is fluvial (streams), marine, or eolian (desert). This decision is based on information provided by cores, geophysical surveys, and other information; not the computer. And never will a geologist rely upon an unedited contour map developed through a kriging or nearest-neighbor algorithm. This is particularly true when the geologist intuits there are other forces at work, such as tectonic faults.

Instead, rather extensive editing takes place, with the geologist applying changes to the rendition using the computer mouse much like a pencil and eraser. And once this computer model is complete, the resulting 2-D contours will still not be really *visualized* as a series of parallel lines. Rather they will be seen as 3-D hills and valleys that stand out in the imagination. These geologists intrinsically understand that 2-D representations help communicate the model, and as such allow interactive visualization. But the true 3-D or 4-D model can only be developed from the human thought process.

As humans cannot read the thoughts of others, however, the geologist must somehow communicate their knowledge to decision makers in the clearest way possible. This must be done by paying close attention to the level of expertise these individuals may have. An executive who rose up through the ranks from the marketing side, for example, may not be educated in geologic map reading. This executive may therefore require nontraditional means of presentation, such as illustrated cross sections or 3-D views developed through surface shading and fractals.[4] Thus one must never lose sight of how to present knowledge developed through the process of visual thinking.

On the other hand, *visual communication can precede visual thinking*. In many cases it directs the model development process in an opposite manner described above. Raw data related to the locations of some spatially variable statistic (i.e., oil characteristics across a petroleum province, the distribution of age groups living within a county) may start with very little informational content.[5] But upon completion of a spatial analysis, the GIS output will attach meaning and structure to the data. Thus trends can be visualized through the production of filtered tables, charts, and maps.

As such, computer analysis has one clear advantage over human thinking. It can distill columns and rows of statistical data more efficiently than a human. A spatial statistical analysis utilizing U.S. TIGER Census data and a corporate database, for example, will make better sense of the data, via visual communication, rather than looking for patterns among tabular sets of numbers through visual thinking.

Hence, visual thinking and visual communication must go hand in hand. Each plays a weighted role in model development and representation, depending on the type of data and the nature of the analysis. Out of this interaction, however, problems will arise. As one proceeds to solve a problem, for example, one develops a familiarity acquired by working through data and information. This familiarity can subjectively blind the analyst to the needs of the audience. The presenter sometimes can forget that the ultimate goal is to convince one's peers and supervisors of the analysis. So what can be done to overcome this problem and better communicate the visual model? Plenty.

Map Types

It should be reiterated that little of the knowledge concerning map design has found its way into computer mapping systems.

—Christopher Jones

Although GIS allows one to produce countless map views and support themes, the respective number of each must remain tightly controlled to produce clear communiqués. Suppose, for instance, that one wished to take a look at the U.S. Gulf of Mexico oil and gas industry in one view. If all of its producing wells, platforms, pipelines, leases, and blocks were included, the representation would become cluttered (see Fig. 8–3).

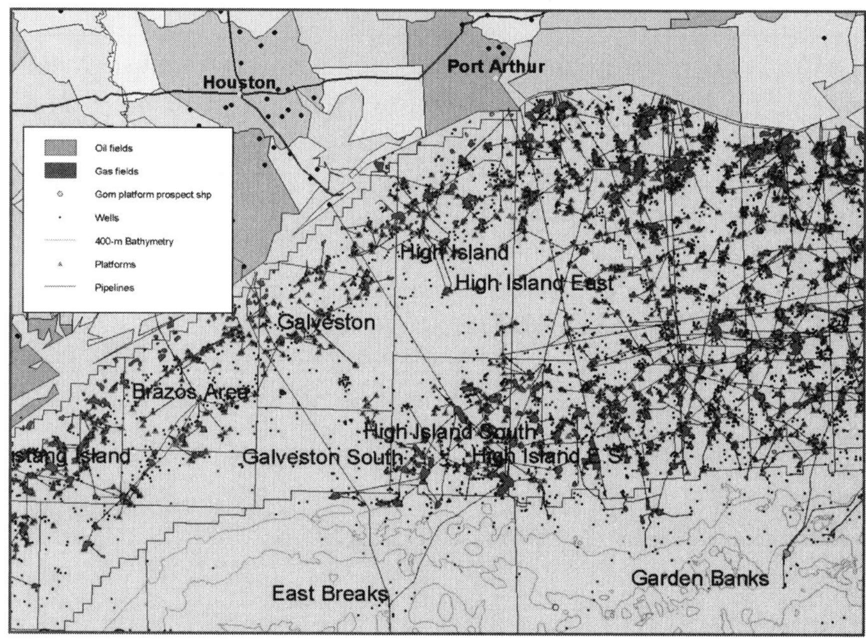

Fig. 8–3 A Busy Map (source: data from U.S. Mineral Management Service, Geoscience Earth & Marine Services Inc., ESRI)

Emplace text on top of these entities, such as well names, pipeline diameters, and so forth, and the presentation becomes absolutely useless. In this situation, many of the themes should be transferred to other views, structured in accordance with the desired topic.

General reference maps

GIS users mostly work with two classes of maps: *general reference* and *thematic*. General reference maps provide a useful means of showing the general assets or activities of a company, say within a worldwide, regional, or territorial perspective. Such maps often include a variety of features, such as political boundaries, physiography (lakes, mountains), and general oilfield infrastructure. They provide a useful means for introducing a presentation or describing a general topic. Basic cartographic applications of abstraction and generalization should be applied to these kinds of maps. Due to the general objective of such displays, however, greater liberty can be taken to make them eye appealing at the expense of exaggeration and geodetic correctness. This is acceptable as long as the audience understands the displays are not to be used in any type of analytical setting.

Thematic maps (map views)

Thematic maps, on the other hand, exclusively focus on one subject or topic. This can include:

- A spotlight on the distribution of a single attribute like oil production (one base layer)
- The integration of several layers in a spatial analysis (location evaluation)
- A mathematical relationship between subdivisions of some feature such as gas composition (i.e., the ratio of methane to total gas)
- A subdivision of distinctly different features of the same category, such as transportation routes (railroads, pipelines, trucks, and tankers)

As such, thematic maps can provide information on a wide variety of subjects ranging from geologic rock type to supply-and-demand relationships. The GIS analyst must always make sure that thematic views show the correct distribution among the various attributes. And when a GIS user designs a thematic map, the data on the map view must be portrayed clearly using sound principles of generalization.

When building a thematic map view, several themes (layers) can be included. However, it is important to limit the number to remain focused. Think through the selection process of these themes beforehand. A rule of thumb is to define the main topic of the individual map view first. Then make a list of themes that should be placed on it. For example, a marketing map should include transportation routes and sales destinations, but not oil wells or geologic contours. If it is important to show such information, dedicated map views should be built for this purpose.

Map view structure

In most GIS programs, a map view can be broken down into the *geographic data frame*, *cartographic elements*, and *elements of style*.

Figure 8–4 (Plate 8–1) shows the elements that can be combined to create a map view.

Geographic data frame. The geographic data frame, or extent of coverage, consists of a specific geographic coverage that is graphically or manually defined in the map view. Direct graphical interactions with data points, lines, or polygons take place within this extent. The geographic data frame is built from themes (layers) and is scale dependent.

Map scale is the ratio of the data frame's extent as compared to the same geographic extent in the physical world. It controls how much area the user will see in the geographic data frame. The user can employ coordinate text inputs to exactly define the frame of coverage. Then the scale can be fine-tuned to some even integer (i.e., manually change scale from 1:47,345 to 1:50,000). Or the user can graphically apply a zoom and pan mode to define the extent.

The data frame is built upon one or more layers that are stacked on top of each other. Each is graphically constrained within the map view.[6] These layers, sometimes called *themes*, are the basic units of geographic representation within the frame of coverage. Some examples include wells, pipelines, and license boundaries. The geographic data frame and the way layers are visually represented can also be controlled by geodetic parameters, such as map projection. In many GIS programs, for example, it is quite easy to change from a geographic latitude and longitude system to a projected grid.

Cartographic elements. Cartographic elements dynamically change when alterations are made to themes and cartographic choices (i.e., scale or projections). Cartographic elements can include legends, scale bars, north arrows, and certain style elements. They are directly linked to the geographic data frame. These elements will change dynamically as numeric classifications, categories, scale, and projections are applied.

Elements of style. Elements of style, which include free-form text, markers, geometric drawings, and pictures, are not directly linked to the geographic data frame. As such, they can be independently applied and have no explicit association with the data frame. Nevertheless, they definitely add information to the cartographic element and data view.

Although map design has not been automated, steps can be taken to produce the best map views for the job.

Presentation Development

To create a presentation that clearly represents an analysis, promotes a recommendation, or supports a decision, the user should go through a serial process. The process should consider the following (see Fig. 8–5):

- Subject. Opportunity, question, and challenge
- Audience. Decision makers, peers, subordinates, and the public
- Material. Data, information, and knowledge converted to map views, themes, categories, and classes
- Style. Abstraction, generalization, symbolization, and elements

Essentially, this process begins with the big picture, funneling down to ever-smaller details.

Presentation Development Steps

Step 1

Define the topic, subject, challenge, or question

Step 2

a) Describe the audience
b) Determine their expertise and level of knowledge in subject matter
c) Use this information to outline presentation

Step 3

a) Organize and prepare themes into map views for target audience based on prior analyses
b) Classify themes into subgroups as needed

Step 4

a) Apply appropriate projection and scales
b) Apply publication style to map and map elements based on sound cartographic techniques of abstratction and generalization

Fig. 8–5 Presentation Organization Steps

The analyst or preparer must first clearly define the subject or topic. For example, should company resources be directed to drill a wildcat well, build a pipeline, or perhaps enter a retail gasoline market? Second, the presenter must think about who will be in attendance, what information should be provided, and the kinds of questions the attendees will ask. For example, will an audit reserve committee be attending, or a group of potential investors? Third, the presenter must determine the kind of views, support themes, classifications, and categories to educate the audience, support the recommendation, or answer questions. Finally, the analyst must build these views with the appropriate scale, projection, generalization, and style. These are necessary so that all of the posted entities and pertinent information will be clearly understood.

Topic and audience

First we should turn our attention to the materials that should be placed in the map views, to the exclusion of graphs and tables. We will address this in the context of the hypothetical Nevada opportunity previously noted in Figure 8–2. We begin by defining the subject and thinking about the audience:

- Subject. Should the company invest in a drilling program in Railroad Valley?
- Audience. Vice president of exploitation, vice president of marketing, and vice president of operations.

Obviously the topic is well defined and provides a starting point for development of the presentation. Each of the executives, however, will have a different set of questions appropriate to each office represented. Moreover, they will also have many questions that overlap other areas of responsibility. The analyst in this case would have to determine the types of map views to make a strong case. Will the information provided from Figure 8–2 be enough, for example, or will management require more? The answer would be more, but organized in a manner such that there is not information overload or confusion.

For example, the exploitation vice president would most likely have questions about reserves and production rates. The marketing vice president will in turn ask about hydrocarbon quality, transportation routes and costs, buyers, and commodity prices. And the operations vice president would most likely be interested in drilling, completion, and production issues. There would be countless more questions, and as such, the analyst will have to think through all of these thoroughly.

Regardless, when considering which views to present to executives, it is important to remember that such an audience will have limited time. As such, the analyst may choose to present the following:

1. Figure 8–2, showing all of the wells drilled in the area but marked with text notes denoting initial production (IP), current production, and cumulative production
2. Cumulative net pay thickness map for all zones marked with total proved and probable reserves
3. Marketing infrastructure map showing field location, ownership, transportation options, storage facilities, and sales destinations

As such, all three map views would provide various degrees of information for the three executives without information overload. View 1 would qualitatively define the drilling risk in relation to the production reward potential. In other words, the number of dry and abandoned wells can be visually assessed against those that are producing. This view would also denote where the best trends are situated. View 2 would then be used to assess resource potential by showing general areal distribution and thickness of all the pay zones. Finally, view 3 would qualify sales opportunities and market routes.

Unfortunately, these static views would leave a lot of unanswered questions for any inquisitive group of executives. Many of you reading this book, for example, may prefer a different set of views than those suggested here. This outlines the greatest shortcoming of paper: map presentations are noninteractive. New information cannot be added nor can old data be edited without going back to the drawing board.

(On a final note, one should take steps to avoid the problem of subjectivity that may interfere with the development of a good presentation. The presenter should show the presentation beforehand to those who have enough knowledge to lend guidance.)

Another way

On the other hand, an analyst could present the information with GIS software, a laptop, and a projector. Then the presenter can readily change the views and even portray the results of the spatial analysis live. Add a printer or e-mail connection, and these can be provided to attendees by a hardcopy or attachment before the discussion even ends.

Let's begin with a dynamic presentation based on view 1 above, for example. If the operations vice president wanted to see how deep the wells need to be drilled across the trend, the presenter could quickly label each well with respective total depth (TD) values. Or for view 2, perhaps the exploitation manager wished to become more familiar with the *sweet spots* along the trend. Instead of focusing on the amalgamated resource base, individual formation themes based on subsea structure and thickness could be interactively turned on and off. Furthermore, subsequent queries could be run on reserves for individual seams.

Finally with view 3, assume the marketing vice president wants to take a look at transportation costs. If the analyst has embedded network impedance values representing tariffs and weather-related downtime risk in the GIS model, competing transportation routes could be quickly compared (see chapter 3). Moreover, buffers and distance measurements could be run to show who owns properties along a given corridor. Or they could be used to measure the distance between the field and sales destinations.

All of this newly generated information could then be easily formatted in the form of text, colors, graphs, and so forth. Selection of what is placed on these views then simply becomes a function of thematic existence or nonexistence, utility and nonutility. It also becomes a function of the formatting capabilities of the GIS.

Style

Maps like stories, have a main theme, point of view, plot, and style.

—David Greenwood

Once the selection process has taken place, all specific themes have to be differentiated by subduing low-priority items and highlighting high-priority items. In practice, this involves an exercise in graphical engineering that can be approached through the proper application of style. *Style* is a function of artistic, graphical, and textual expression. It includes everything from the selection of text placement adjacent to an entity to polygon fill patterns used to distinguish one category from another. Its application requires numerous choices, often involving minute details. Style, however, can be tackled in a structured manner by understanding the various components that comprise it.

These building blocks of style, which are based on the concept of abstraction and generalization, consist of:

- Numeric classification
- Nonnumeric categorization
- Simplification
- Exaggeration
- Displacement
- Enhancement
- Transformation
- Symbolization
- Deduction

Representations

There are four primary representations of mapped quantities. These include:

- Discrete features dominated by vector entities (but also rasters)
- Images and sampled grids (remote sensing)
- Surfaces or continuous phenomena (TINs)
- Data summed by area (statistical representations)[7, 8]

As one gains experience with GIS, it will be found that the type of attribute one wishes to portray, model, or visualize will form the basis for symbolic design. Numeric data, for example, can be effectively represented with symbols that change size or color according to attribute value [see Fig. 2–7 (Plate 2–5)]. On the other hand, nonnumeric attributes that categorize a particular group of features may be represented best with unique symbols for each feature.

Numeric classification

With numeric classification, the goal is to scale and group features by their attributes. In other words, classification strives to express the distribution of some group of attributes in a logical manner. Recalling chapter 3, the more common classification groups include equal interval, quantile, natural breaks, and standard deviation.

To improve communication, objects can be grouped into classes. For example, oil wells located across a large field will have a wide disparity of production rates. Suppose one wished to uniquely depict a large number of wells with specific production rates sized proportionately to one another. The theme would turn out to be incomprehensible. Thus, it becomes best to classify groupings of symbol sizes into production ranges, such as 0–10 BOPD, 11–100 BOPD, and 101–1000 BOPD.

With one legend symbol representing many wells, this provides a general idea about production. It also avoids unneeded specificity and a multitude of symbol sizes. Remember, the human mind can visually distinguish up to seven ranges of size, color, or shades with a high degree of cognitive discernment. More than this, and it becomes confusing.

There are at least five ways of presenting numeric classifications. These are: gradational colors, unique colors, predefined symbol sizes, proportional symbol sizes, and some combination thereof.[9]

Gradational color. To apply a scheme of gradational color, the classification of attributes can be normalized by subdividing a collection of values. They are then classified through some statistical means (i.e., linear, logarithmic, or standard deviation). The mathematical outputs of these divisional calculations can then be discerned from one another by altering the hue of an individual color. In most cases, value differentiation begins with a light color, then blends increasingly darker. This drawing method provides an effective means for presenting numeric data that are gradational or continuous in nature. Examples of such data include temperature, pressure, elevation, or resource distributions.[10] This method also provides an effective way to treat area-based themes.

The hue of a color can also be altered within a contiguous distribution. Bathymetric maps, for example, can portray deepwater areas as dark blue up to the shelf. Lighter shades of blue can be applied as the water becomes shallower toward the shoreline.

Unique colors. The other obvious choice in applying color utilizes unique colors to distinguish classes. This is ill advised, however, as variegated color schemes tend to be visually confusing.[11]

Predefined symbol sizes. With predefined symbol sizes, the user specifies a value field, an optional normalization field, and a classification.[12] The range of values is first subdivided into classes based on the most appropriate classification (natural breaks, etc.). Next a base symbol is chosen. Finally varying symbol sizes are applied, with small symbols typically representing small values and large symbols representing large values.

This drawing method is suitable for numeric data that represent a rank or progression of values. Care must be taken to ensure that the symbol size selection does not allow larger value entities to overlap smaller value entities. The symbols can be further distinguished by changing the color of the symbol body and even the outline.

PROPORTIONAL SYMBOL SIZES. The application of proportional symbols uses a similar process to graduated symbols. However, there is no classification of values, with the size of each symbol varying in proportion to the attribute value.[13] The procedure works by specifying a value field, then an optional normalization field, and finally the units for display. Attribute values can be varied by symbol size, area, or radius. Again, large symbols should be drawn first to avoid the masking of smaller symbols.

COMBINATIONS. In some cases, a variety of combinations listed above can be used.

Nonnumeric categorization

Categories should not be confused with numeric classification. Whereas the latter deals mostly with purely numeric or statistical data, categories deal with nonnumeric attributes. A category, for example, can represent the ranking of citizens by race; or it may show a subdivision of surface geology into eras, periods, and epochs.

When the analyst classifies features by type, each feature must have a code that identifies its type.[14] For example:

- Paleozoic era = 100
- Mesozoic era = 200
- Cenozoic era = 300

Furthermore, many categories are hierarchical, so that major types can be subdivided into subtypes. For example, Mesozoic rocks can be further categorized by the following periods and respective codes:

- Triassic = 201
- Jurassic = 202
- Cretaceous = 203

With GIS, it then becomes possible to categorize and subcategorize these as desired using the style points noted above. There can be a single unit, such as Mesozoic, based on the code series of 200. Or there can be subunits (Triassic, Jurassic, and Cretaceous) based on subcodes 201–203. Therefore, to distinguish one era from the next, a unique color for each era can be applied. To distinguish all subunits (periods) of a particular category (era) as a single unit, each period would retain the same color.

Different hues, symbol types, gradational fills, or hatch types, nevertheless, can be applied as the codes allow it. Table 8–1 shows how relational tables make it possible to apply different style patterns to categorize and subcategorize features as long as the codes allow it (note numbers 201–203 in lower table).

Era	Code	Area Color
Cenozoic	300	Yellow
Mesozoic	200	Green
Paleozoic	100	Red

Period	Code	ID	Area Color	Fill Style
Cretaceous	200	201	Green	Dot pattern
Jurassic	200	202	Green	Vertical Lines
Triassic	200	203	Green	Cross pattern

Table 8–1 Relational Tables for Style Patterns

For example, a geologic surface map with GIS-coded attributes could contain formations from both the Paleozoic and Mesozoic eras. In addition it could also include all of the periods from Cambrian through Cretaceous. Each period could readily be associated with its governing era by applying a similar color to all periods within the same era (Paleozoic or Mesozoic). Next each period can be distinguished by applying a different hue of the same color or a different hatch pattern [see Fig. 8–6 (Plate 8–2)].

In comparison with numeric classification, different color schemes can and should be applied to distinguish thematic categories. Figure 8–6 (Plate 8–2) shows how this concept is applied to a Russian surface geology map. In this figure, Cenozoic rocks are yellow, Mesozoic rocks are green, and Paleozoic rocks are red. To distinguish periods within these eras, different fill patterns can be applied.

Thus mapping entities and features by category provides information and reveals patterns on interactions between assets, resources, and features. Mapping all pipelines with a single black line, for example, indicates location and little else. Gas pipelines, on the other hand, can be distinguished from the oil pipelines by using two different colors. Junctions and flow direction also can be denoted with unique symbols and arrows. As a result, one can begin to evaluate tie-in and transportation options.

As noted, if there are numerous categories that are too complex to distinguish through variations in color, symbol type, and so forth, it is best to build another map view. For example, consider a given area with numerous oil and gas pipelines crisscrossed with tank farms and compressor stations. In this instance, separate oil and gas pipeline map views should be created. Furthermore, this provides more room for adding informative text, such as owner name and pipeline diameter. A single theme or layer should never show too many categories, or else the view will become confusing.[15]

Simplification

With simplification, the analyst or presenter selects the most important attributes and feature and hides any unwanted details. Because symbols take up room on a view or map, it becomes clear that as the map scale decreases in size, fewer features can be represented. With GIS, part of the solution is to deselect unnecessary features. In some cases this is not enough, however, and the view must be refined through *smoothing*, *amalgamation*, and *conceptualization*.

SMOOTHING. Smoothing can be applied to certain entities like rivers and coastlines, where each curve cannot be depicted due to scale or because the detail would clutter the map. In such cases, a traditional mapmaker would smooth out the meanders by removing every *n*th node along the line. Alternately, the mapmaker could run a Douglas algorithm that retains *extreme* points along a curve.[16] With GIS, however, smoothing intrinsically occurs with dynamic changes in scale (see Fig. 3–1). As such smoothing is usually not deemed a necessity by the author.

AMALGAMATION. Amalgamation, on the other hand, can be used to merge originally distinct or entirely separate objects into one representative object.[17] This function can be applied to numerous features in a small area. One suitable situation, for example, occurs when a large number of small oil fields appears as a crowd of points. To simplify matters, these entities can be grouped (compacted) into one large oil province denoted by a single large dot. An outline buffer can also be run along the external extents of the outermost oil fields to show the outermost perimeter. With GIS, this provides an excellent option and should be considered when warranted.

Conceptualization. Another form of simplification, conceptualization, has to do with the need to geometrically represent an object. The goal is to include enough detail so that it can be seen for what it is, but avoid adding unnecessary or hard-to-portray detail. For instance, it would be impractical to portray a refinery with all of its processing vessels, towers, piping, and components, even on a map scale of 1:5000.

Yet the facility could be well represented by typifying the major activity centers of distillation, cracking, hydrotreating, alkalization, and coking into readily distinguishable zones. These can be denoted through colors, shading, or uniquely compartmentalized shapes. With GIS, it is then possible to link dynamically these areas to the actual plant design files (i.e., CAD drawings). Consequently they can be pulled up on demand or emplaced within the map view as embedded images.

Exaggeration

With exaggeration, the map preparer enhances or emphasizes important attribute characteristics. Only on very large scale views, such as 1:2000, can features such as pipelines and surface facilities be shown in detail without enlarging size. Due to the effects of scale, for example, certain features must be shown larger than their actual relative size, or else they will be invisible. Thus, entities such as pipelines and roads must be scaled up from their true correct map scale size to show their relationship with other features.

Displacement

When symbols and annotative text begin to compete for space, it often becomes a choice of displacement or elimination. According to Jones, "The need for displacement arises because adjacent generalized geometric representations or graphical symbols may overlap each other or become too close to be clearly discernible."[18] One must remember that the true geographic position will not be shown correctly in this case.

Elimination

Elimination, on the other hand, becomes a choice of which entity or element is the most important to be shown on the map view. Sometimes the entity or layer to be deleted is considered to be important. In these cases the presenter can layer the largest entity underneath the smaller, and fill it with a different color or pattern than the smaller. Or the user can choose to reduce the size of the competing entity in order to retain it.

Transformation

With transformation, the user converts one type of conceptual or representational entity to another. Consider changing the representation of a refinery from a point, with symbol size indicating gasoline output, to an area representing its actual areal outline. This is a transformation from a point to a polygon. Represented as an area, information related to gasoline output is lost, unless some other graphical or text-based variable is used to reapply the information.

Symbolization

Symbolization depends on graphical representations needed to communicate what each object or feature is. Discussed in prior chapters, primary symbols are denoted by point, line, and polygon entities. The GIS user employs these graphic marks to portray entities, concepts, classification of facts, or the characteristics of a geographical distribution. Again, scale plays a decisive role in determining which kind of entity (hence symbol type) will be utilized. An island, for example, will be viewed as a point entity on a small-scale map. However, it will be represented as an area on a large-scale map.

Specific entities or features chosen may be:

1. Representational—a derrick depicting an active drilling rig
2. Conventional—a star representing a state capital
3. Arbitrary—a screw symbol representing a progressing-cavity pump (PCP)

There are cases where representations are universally known or commonly applied within a particular industry. In these cases symbol denotation may not be required on the legend, depending on the familiarity of the audience with the symbols. When an arbitrary symbol system is used, however, the noted object or feature must always be denoted.

There are four basic types of symbols used to present descriptive information about point, line, and area fill features. These are: *marker symbols*, *line symbols*, *fill symbols*, and *text symbols* (see Fig. 8–7).[19]

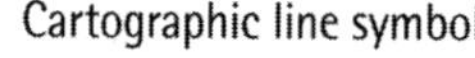
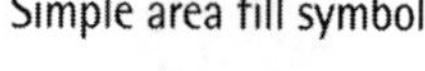

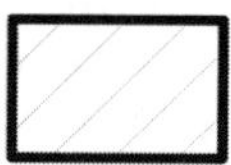

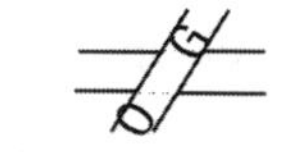

Fig. 8–7 Symbol Types

Marker symbols. Marker symbols can be based on a single character such as a star representing a state capital. Other marker symbols might be an arrow representing flow direction or a thumbprint picture portraying a specific subject or denoting a file link. A multifaceted symbol also might be used that includes a combination of shape and text, such as a highway shield.

Line symbols. Line symbols can be composed of various widths, colors, dashes, hatches, markers, arrowheads, and so forth. For example, a dirt road can be differentiated from a major freeway by dashing the line and making it light gray.

Fill symbols. Fill symbols involve polygonal features and can consist of simple line, marker, gradient, and picture fills, or some combination thereof.[20] A simple fill uses the properties of color, outline style, and outline width, and fill patterns such as crosshatches or vertical hatches. A line fill is more complex than a simple fill, as it can use patterns that involve straight or curvilinear representations at any angle and separation. A marker fill uses a grid of marker symbols that can be spaced and rotated. A gradient fill uses a blend of two colors that transition from one side of the polygon to the other.

Gradient fills can blend color in one direction, radially blend in a circular manner, or geometrically blend from the center outward. Gradient fills also can buffer according to some percentage value based on predefined limits. For example, dark blue signifies deep ocean and various shades of light blue are applied along the shelf.

Text symbols. Text symbols often consist of labels placed next to, along, or inside an entity. There are two types of annotation: feature-linked annotation and free-form label.[21] Feature-linked annotation is a natural function of most GIS programs, as it provides dynamic labeling of features. Typically, any attribute field within the database that is applied to an entity can be annotated next to its appropriate entity. For example, state outlines represented by polygons can be easily labeled with the name of the state.

Free-form annotation, on the other hand, is not bound by any attribute-entity relationships. Instead, text can be placed anywhere in the geographic data frame. Thus notes, titles, additional information, and so forth can be added directly to the map view.

For both feature-linked and free-form annotation, most GIS programs provide a variety of fonts, text sizes, and color options. Additionally, scalable options can be applied to visibly maintain the text as scale changes are introduced.

Deduction

Map design cannot be automated. Rather it requires skills in developing the proper style to best represent data and information in many cases that are unique. A map user, with personal information about a particular topic, can further deduce insights. These thoughts then can be amplified on the map through combinations of style noted above. For example, statistical data acquired through a survey can only be qualified by the survey taker. If that individual happens to be the GIS analyst, certain nuances and environmental factors can be denoted, annotated, or factored into a classification.

For example, a surveyor might question a population about a new type of soft drink throughout the course of a year. From this information, however, the surveyor will likely infer that during winter, their answers will be quite different than if the same individuals were questioned in the summer (i.e., less thirsty). Consequently, each precinct could be color coded according to the season of data collection. Then the statistical data could be introduced as point symbols. In this way, additional insights can be provided to the analysis.

This concept of course can be extended into petroleum. A geologist who personally examined a wellbore core, for example, could make insightful judgments as to core quality. For example, a driller pulling the core barrel may note possible contamination from a cracked liner. Or the core quality may vary from rubble to consolidated material. These insights could then be qualified on an ordinal scale from unconsolidated to consolidated, or from contaminated to intact.

Another example might occur with an experienced geologist working on a regional map that tectonically distinguishes one basin from another. The way this information can be added depends on one's experience in dealing with style. Obviously, the various categories that make up generalization can be cross-matched, which is one of the reasons why map presentation cannot be automated. Instead this requires the human touch.

Map Scale

The previous discussion on generalization and symbolization points out one important observation. Cartographic scale and data resolution are the determining factors behind map design. Thus, any user of GIS should become familiar with the basic treatment of these subjects.

Scale types

There are three types that must be understood: *English verbal*, *representative fraction*, and *graphic*.[22]

English verbal. The English verbal (EV) scale is probably the most difficult to master, yet once it is understand, simple algebraic conversions can be used to perform spatial investigations. Think about the following: "1 in. equals 5 mi." The problem here lies in learning how to deal with different unit types on opposite sides of the statement (see Box 8–1).

BOX 8-1:
Unit Conversions

1 mile = 5,280 feet = 63,360 inches
1 kilometer = 1,000 meters = 100,000 centimeters
1 mile = 1.61 kilometers
1 inch = 2.54 centimeters (exactly)
1 square mile = 640 acres = 259 hectares
1 hectare = 2.471 acres = 10,000 square meters
16 hectares = 160,000 square meters = 39.53 acres (or about 40 acres)

For example, there are 12 in./ft, and 5280 ft/mi. One could measure off 3 in. on a paper map, based an ER scale of 1 in. equals 5 mi. Next a fractional computation must be conducted to find the true distance, whether it is with a calculator or in a GIS. Cartographers of old had to deal with unit conversions regularly in order to obtain distance measurements using a ruler. Yet GIS eliminates this tedious, sometimes error-prone chore through an integral measurement tool. This can be applied graphically or manually to measure distances between two or more points (see Fig. 8–1). However, if paper outputs of a GIS analysis are provided, then it is still important to be able to apply algebraic routines to come up with an answer.

Representative fractions. As compared to EV, representative fractions (RF) provide a more straightforward way of calculating distance. On almost all topographic maps produced in Russia, for example, one will see an expression like 1:10,000. What this means is for that particular map, one map unit equals 10,000 of the same units on the Earth. The unit type is not important, as long as the unit types on both sides of the ratio are the same. Thus, if the RF of a map is 1:10,000, this means that 1 in. on the map equals 10,000 in. on the Earth; or 1 cm equals 10,000 cm, depending on the measurement system.

Graphic scale. The graphic scale (GS) can be likened to printing a small ruler within the margin of the map.[23] The divisions denoted within this "ruler," however, do not always correspond to some established length measurement like centimeters or inches. This is especially true with non-GIS-built maps. Instead, some printed unit length, for instance 1 km, may be subdivided into any number of sections that may or may not coincide with some measurement system. Thus someone measuring with an engineering ruler on a printed map using a GS will find that the unit divisions most likely do not match. On the other hand, if the proper projection is applied, the "ruler" and subdivisions given within will accurately represent some definitive distance on the ground.

One note of importance: it is always best to use a printed graphic bar scale on a published hardcopy map rather than a text-based RF or EV scale. The map scale will be correct for the original map. If someone decided to copy a map, however, and inadvertently or purposely reduced the map in size, the RF scale would no longer hold true. However, a printed bar scale would remain true.

Resolution

Resolution can be thought of as the ability to discern detail. Although reliant on map scale for visual portrayals, resolution depends on the informational intensity of a map. It does not depend on the mathematical measurement of some ground distance to its geographical extent in the map view.[24] Obviously, more detail can be visualized on a large-scale map than on a small-scale map. However, if the informational resolution is low, the value of the GIS outputs will be reduced in quality. If a GIS captures more detail, however, the analysis will be more accurate.

In his book, *Spatial Reasoning for Effective GIS*, Joseph Berry has defined four types of resolution:

1. Minimum mapping resolution (MMR)
2. Spatial resolution
3. Thematic resolution
4. Temporal resolution

Minimum mapping resolution

Minimum mapping resolution identifies the smallest physical grouping of a map theme, or the level of spatial aggregation. This can be thought of as the smallest area that can be circled and called a single group. One example would be the number of wells in a distinguishable cluster that can be readily delineated as an oilfield. Smaller subgroupings obviously produce a higher MMR.

Spatial resolution

Spatial resolution refers to the size of the smallest object or ground feature that can be distinguished in an image or vector system. In other words, it identifies the smallest addressable unit of space. In the vector and raster systems, the main units of spatial address consist of the line segment and pixel, respectively. For vectors, spatial resolution is a function of the density of points along the line. The smaller the distance between each point, the higher the spatial resolution. With rasters, spatial resolution is a function of cell size—the smaller the cell, the higher the spatial resolution.

Thematic resolution

Thematic resolution identifies the smallest grouping of categories or classes in a map theme. For example, a geologic map can be subdivided by age, rock type, tectonic regime, and so forth. The further a map theme can be subdivided, the higher the thematic resolution.

Temporal resolution

Temporal resolution identifies the frequency of a map update. For example, a landman working for a small oil company may only require concession maps with lease ownership information updated yearly. In contrast, a large company representative may require enterprise-wide updates monthly or continuously. The more often the map is updated, the higher the temporal resolution.

3-D Presentations

The next subjects of discussion refer to two groups of analysis and presentation, notably 3-D and 4-D (time-lapse) applications. These applications have greatly expanded the realm of interactive communication and visual thinking. A presentation in 3-D can be divided into two groups. The first represents those that simulate the appearance of the third dimension. (This is sometimes called the two and one-half dimensional approach.)[25] The second group provides true graphical interaction.[26] For those programs that do not provide true 3-D capabilities, there are at least three ways in which GIS can portray relief on a 2-D computer screen or printout.

GIS relief portrayal

Hillshading. The first method consists of hillshading, a technique that infers the position of the Sun in the sky and then calculates the angle between the direction to the Sun and the perpendicular to the surface.[27] This angle is used to apply various shades of brightness, which is proportional to the cosine of the angle between the surface perpendicular and the light source vector, to a surface. Hillshading produces realistic views of the terrain and works well with TINs.

Elevation with graduated colors. The second method, the simulation of elevation through the use of graduated colors, can also be applied with TINs. However, it uses various hues of color or colors to illustrate a range of elevations. This technique interpolates contour lines for each TIN face, after which the user applies the desired graduated interval.

Slope with graduated colors. The third method, slope depiction using graduated colors, calculates the angle between the perpendicular of a surface and the plane of the Earth. This method includes the application of a graduated range of colors and hues applied to angles between 0° and 90°.[28]

True 3-D

Noninteractive applications, like those just mentioned, do not provide the necessary foundation from which to evaluate and manage petroleum resources. This is in many cases not useful for the petroleum professionals, especially those involved in geology and subsurface engineering. Consequently, an entire cottage industry has developed competing 3-D software packages. Such products include Geographix (Landmark Graphics), Eclipse (Schlumberger), Seisworks (Landmark), and Surfer (Golden Software). The development of these software products has helped geoscientists and engineers model the Earth. These advanced modeling packages can be called GIS by some definitions. This is because they provide spatially dependent geographic and interactive information that can be queried, culled, and analyzed.

On the other hand, these programs are primarily dedicated to upstream activities, mostly dominated by the exploration and production (E&P) sector. As such they exclude enterprise-wide GIS utilities for marketing and business applications. Thus, there is obviously a need for the infusion of such advancements into other business endeavors.

Along this line of thinking, some authors have considered transforming zone-based census and point-based address data into an information structure that represents population distribution as a continuous surface. Such an approach, however, "requires a change in the basic assumption of data representation, namely that the data value is seen not as a property of an area or a point, but rather property of a location at which the population are distributed."[29] Obvious implications include new ways in which to evaluate supply-and-demand issues, sales opportunities, operations, production, and so forth.

In the search to advance the use of true 3-D interaction, some industry GIS specialists have begun to adapt off-the-shelf programs to fit a need. John Grace with Earth Science Associates, for example, has developed such a utility. His program involves the use of ESRI's desktop GIS—ArcGIS—and its respective add-ins, Spatial Analyst and 3-D Analyst. These are integrated with Oracle, a relational database manager (see chapter 3).

Working with publicly available Gulf of Mexico (GOM) data provided by the U.S. Minerals Management Service (MMS), Grace has developed a means to stack and integrate data. This works by integrating structure contours, directional well surveys, zonal production, log curves, and even fossil data in a variety of ways to show what is taking place in the subsurface.

This is demonstrated in Figure 8–8 (Plate 8–3), where Grace has generated a structural contour map on top of the Bul-1 reservoir of the Ewing Bank 873 field, in this case published as a 2-D view. By using GIS symbolization tools, however, Grace has applied a pie-chart utility to represent cumulative production through 2000. As such, the relative size of each pie indicates production. Furthermore, he layered in the directional well profiles to show their emplacement within the structure of the top of the Bul-1 reservoir.

At first glance, the presentation in Figure 8–8a (Plate 8–3a) appears to be a 2-D rendition. Nothing is further from the truth, however, as the view in the GIS program can be rotated, zoomed, and panned in all directions [see Fig. 8–8b (Plate 8–3b)]. Moreover, all of the data available in the database can be queried from the map view, through graphical or text-based means. As such, the GIS can interactively be used to gather other information, such as reservoir age or well geometries. This 3-D view of the Bul-1 reservoir shows a more visual portrayal of the same structure that otherwise can only be visually imagined by a trained geologist from the 2-D contours shown in Figure 8–8a (Plate 8–3a).

Another 3-D example by Earth Science Associates is illustrated in Figure 8–9 (Plate 8–4), which shows the High Island 474A field in western GOM. This field consists of several Pleistocene turbidite (marine landslides) and fan complexes that have been deposited from northwest to southeast. In this 3-D view, the viewer (camera) looks along the direction of deposition at an altitude above sea level. Sediments are shown originating at the top-back of the field model, flowing down dip in the direction of the viewer. The scene furthermore shows the graphical cage of the MMS block boundaries for geographic reference, with well profiles represented by 3-D directional surveys (black lines).

Applying sound presentation practice, Grace distinguishes each unique sand with a different color. Thus it is easy to see where the separate sands are and where they are intercalated. And because the model has been developed in a 3-D GIS environment, other views can be rendered with any attributes capable of being mapped. These include sand colors reflecting gas-oil ratios, estimated ultimate recovery, porosity, and even thickness.

Time-Lapse Animation

In prior chapters, discussions on time-lapse animations spoke of the great value 4-D GIS can add to spatial interpretations. For example, space-time analyses utilizing repetitive seismic surveys have been successfully applied in mapping permeability heterogeneities across oilfields. They have also been used to keep track of CO_2 floods.[30, 31] The following two examples highlight analyses that have been applied in the downstream and upstream sectors of the petroleum industry.

Downstream

In the downstream sector, it is possible to follow the changing technical and production characteristics of U.S. refineries over time. This analysis can be used for marketing, business development, and even energy policy purposes. According to *Oil & Gas Journal's* Energy Database, for example, the number of U.S. refiners fell from 193 facilities in 1990 to 154 in 2000. Only one state, Louisiana, increased the number of refineries.[32] On the other hand, crude capacity increased from 15.55 to 16.54 million barrels of oil per calendar day (BOPCD).

Unfortunately, diverse standards put into place by various state governments have forced refiners to rapidly upgrade facilities without guidance from any type of national energy policy. This has resulted in segregated markets from state to state that have produced tighter-than-normal relationships between supply and demand. "The timing and size of the necessary refinery and distribution investments to reduce sulfur in gasoline and diesel, eliminate methyl tertiary butyl ether, and make other product specification changes such as reducing toxic emissions from vehicles" was unprecedented in the petroleum industry.[33]

As such, California suffered volatile retail gasoline prices at the turn of the century because regulatory guidelines disallowed the use of other gasolines from other refining centers. Moreover, while refining capacity increased overall across the United States, California's production went into decline. This was a result of costly and compulsory technical upgrades that did not make economic sense for some companies. From 1989 to 1999, for example, California's refiners closed 10 facilities, with crude inputs dropping from 2.2 million to 1.97 million BOPCD.

By plotting this information in a GIS, a geographic picture of production capacity across the state of California can be visualized. In the San Francisco area, for example, 7 refineries were able to handle 844,000 BOPCD in 1989 [see Fig. 8–10 (Plate 8–5)]. By 1999, however, 2 facilities had gone out of business, reducing the refining throughput by 53,700 BOPCD. At the same time, the local population grew by 319,900 residents to 3.387 million, placing heavier demand on supply. In addition, investments in regionally specific reformulated gasoline increased over the 10-year period, placing technical demands on the refiners. With rising energy demand and a declining refining base, problems were bound to arise.

Upstream

Another example of time-lapse analysis has been addressed by Earth Science Associates. Returning to the GOM examples described in the 3-D discussion, Grace has utilized public information to show the utility of analyzing production and test data from a reservoir during a period of time. In Figure 8–11 (Plate 8–6) and Figure 8–12 (Plate 8–7), for example, he gridded and analyzed annual gas and oil production for each well in the Bul-1 reservoir at EW873. These were mapped to show the areal and annual variations in production, in this case biennially.

The maps for each year were then draped on top of the 3-D reservoir, creating a time-lapse view of annual variations in gas and oil production for the reservoir. In this case, the range for gas goes from white to dark red, with dark red indicating highest gas production [see Fig. 8–11 (Plate 8–6)]. In a similar manner, gradational hues of green vary from white to dark green for oil, with dark green indicating the highest annual oil production [see Fig. 8–12 (Plate 8–7)].

Thus it becomes possible to see reservoir depletion through the visualization of color. Grace has taken this analysis one step further by combining data from both gas and oil production to create an animation of annual variations in the gas to oil ratio. This clearly showed the differences in reservoir performance that could not be seen by plotting oil and gas production over time.

Reservoir thickness and structure are already reflected in the geometry of the 3-D polygon, representative of the reservoir itself. Thus draping the time-lapse animation on top puts the annual production maps in a useful context. Changes in production rate, water cut, gas-oil ratio and pressures can then be displayed against reservoir geometry and fault intersections. All of the sandstone beds, wells, and completions in the scene can be queried directly on the screen for underlying attributes. These attributes can cover the rock properties of the sands to the monthly production of each completion. Further refinements can be applied by changing the choice of the gridding algorithm (i.e., kriging, nearest neighbor, etc.).

Moreover, Grace has looked into modeling the value at each point based on the physics of the underlying process being mapped. This approach borrows from reservoir engineering principles. Production is gridded using a radial flow equation that describes the movement of fluids in rock in response to pressure differentials. As such, insights into reservoir compartmentalization, permeability changes, liquid and gas viscosity, and primary and secondary completions can be examined in a space-time context.

Conclusion

The art of presentation must be mastered; otherwise, all of the efforts put into a GIS analysis will fail. First and foremost, a structured methodology that considers the subject, audience, material, and style should be applied in the design process. This is done to persuade the audience of one's recommendations. In other words, all materials should be put together so that information and knowledge are best presented to decision makers, peers, subordinates, and the public.

These materials should be constructed upon sound practices of cartography, generalization, and style. With GIS, moreover, modern software applications provide new technologies for delivering information, most noteworthy of which includes 3-D and 4-D views. These resources should be drawn upon, especially in the petroleum industry, as data often have a Z factor or change with time.

Closing comments

As this is the final chapter in this book, the author hopes that this mission in GIS education has been a success. The broad scope of the subject obviously cannot be entirely broached within a single volume. However, it is this author's sincerest desire that petroleum professionals not familiar with GIS take a look at its potential for improving the decision-making process. The tools are in place; it is only a matter of putting them to work.

References

[1] Johnson, E.H. 1993. *Oil & Gas Journal* 91:20 (May 17).

[2] PennEnergy Data. (www.pennenergydata.com).

[3] *Oil & Gas Journal* Online Research Center (www.ogjresearch.com); Daniel Johnston Nelson Cost Index Analysis 2001 and *Oil & Gas Journal 2001 Annual Worldwide Refining.*

[4] Jones, C. 1997. *Geographical Information Systems and Computer Cartography.* Essex, England: Longman. p. 267.

[5] Ibid.

[6] Zeiler, M. 1999. *Modeling Our World—The ESRI Guide to Geodatabase Design.* Redlands, CA: ESRI p. 27.

[7] Mitchell, A. *The ESRI Guide to GIS Analysis* Redlands, CA:ESRI. p. 40.

[8] Zeiler, p. 25

[9] Ibid. pp. 33–35.

[10] Ibid. p. 34.

[11] Mitchell, p. 59.

[12] Zeiler, p. 33.

[13] Ibid. p. 35.

[14] Mitchell, p. 25.

[15] Ibid. p. 31.

[16] Jones, p. 276.

[17] Ibid. p. 283.

[18] Jones, p. 286.

[19] Zeiler, p. 30.

[20] Ibid. p. 31.

[21] Zeiler, p. 95.

[22] Greenwood, D. 1964. *Mapping.* Chicago and London: The University of Chicago. pp. 43–44.

[23] Ibid. p. 44.

[24] Berry, J.K. 1997. *Spatial Reasoning for Effective GIS.* Fort Collins, CO: GIS World. p. 115.

[25] Heywood, I., S. Cornelius, and S. Carver. 1998. *An Introduction to Geographic Information Systems.* New York: Addison Wesley Longman. p. 63.

26 Ibid.

27 Zeiler, p. 42.

28 Ibid. p. 43.

29 Fotheringham, S. and P. Rogerson. 1994. *Spatial Analysis and GIS*. London: Taylor and Francis. pp. 249–250.

30 Datta-Gupta, A. and D.W. Vasco. 2001. "Production Tomography Merges Geophysics with Reservoir Engineering." *Oil & Gas Journal* (June 4). p. 75.

31 Moritis, G. 2000. "Seismic Surveys Provide Baseline for CO_2 Flood." *Oil & Gas Journal* (June 5). p. 41.

32 *Oil & Gas Journal's* Energy Database.

33 *Oil & Gas Journal Online*. 2000. (June 23).

Appendix

Acronyms and Abbreviations

acre (ac)
Advanced Spaceborne Thermal Emission and Reflection Radiometer (ASTER)
American Community Survey (ACS)
antenna pattern corrections (APC)
Association for Geographic Information (AGI)
automated mapping and facilities management (AM/FM)
barrels of oil per calendar day (BOPCD)
centimeter (cm)
Clouds and the Earth's Radiant Energy System (CERES)
coalbed methane (CBM)
computerized mapping program (CMP)
Continental U.S. (CONUS)
cost, insurance, and freight (c.i.f.)
Critical Infrastructure Protection Initiatives (CIPI-1)
decision support tool (DST)
Defense Information Systems Agency (DISA)
detection and classification tool (DCT)
differential GPS (DGPS)
digital elevation files (DEFs)
digital elevation models (DEMs)
electromagnetic (EM)
English verbal (EV)
Enhanced Thematic Mapper Plus (ETM+)
Environmental Systems Research Institute (ESRI)
European Petroleum Survey Group (EPSG)
exploration and production (E&P)
feet (ft)

former Soviet Union (FSU)
free on board (f.o.b.)
geographic-based file/dual independent map encoding (GBF/DIME)
Geographic Information for Sustainable Development Initial Capability Pilot (GISD-ICP)
geographic information system (GIS)
geographic positioning system (GPS)
Geoscience Earth and Marine Services Inc. (GEMS)
Geospatial Information & Technology Association (GITA)
graphic scale (GS)
hectacre (ha)
inches (in.)
Indian Remote Sensing (IRS) satellite system
infrared (IR)
initial production (IP)
intertial navigation system (INS)
Gulf of Mexico (GOM)
kilometers (km)
micrometer (μm)
miles (mi)
miles per hour (mph)
million cubic feet (mcf)
million cubic feet per day (MMcfd)
minimum mapping resolution (MMR)
Moderate Resolution Imaging Spectroradiometer (MODIS)
monitor control stations (MCS)
Multiangle Imaging Spectroradiometer (MISR)
multispectral scanners (MSS)
nanometer (nm)
National Aeronautics and Space Administration (NASA)
net present value (NPV)
North America Datum of 1927 (NAD27)
parts per million (ppm)
Petrotechnical Open Software Corporation (POSC)
photo scale (PS)

positional dilution of precision (PDOP)
progressing-cavity pump (PCP)
Quality Engineering and Survey Technology Ltd. (Quest)
representative fractions (RF)
satellite laser ranging (SLR)
selective availability (SA)
Shuttle Radar Topography Mission (SRTM)
Space and Naval Warfare Systems Command (SPAWAR)
Spatial Data Transfer Standard (SDTS)
spatial decision support systems (SDSS)
Système Probatoire d'Observation de la Terre (SPOT)
structured query language (SQL)
supervisory control and data acquisition (SCADA)
Texas Legislative Council (TLC)
thematic mapper (TM)
Topologically Integrated Geographic Encoding and Referencing (TIGER)
total depth (TD)
triangulated irregular networks (TIN)
U.S. Department of Defense (DOD)
U.S. Geological Survey (USGS)
U.S. Minerals Management Service (MMS)
ultraviolet (UV)
universal transverse Mercator (UTM)
Urban and Regional Information System Association (URISA)
Very long baseline interferometry (VLBI)
Warsaw Pact Gauss Kruger (GK)
World Geodetic System of 1984 (WGS84)
Wyoming Oil & Gas Conservation Commission (WOGCC)

Index

A

B

C

D

E

F

G

H

I

J

K

L

M

N

O

P

Q

R

S

T

U

V

W–Z